VOLUME XVI
PART III

ANNALS OF THE INTERNATIONAL GEOPHYSICAL YEAR

ANNALS OF THE
INTERNATIONAL GEOPHYSICAL YEAR

1957 *1958*

VOLUME XVI

PART III

CALENDAR RECORD FOR THE INTERNATIONAL GEOPHYSICAL COOPERATION 1959

International Council of Scientific Unions
Comité International de Géophysique
(C I G)

PERGAMON PRESS
Oxford · London · New York · Paris

PERGAMON PRESS LTD.
Headington Hill Hall, Oxford
4 & 5 Fitzroy Square, London W.1

PERGAMON PRESS INC.
122 East 55th Street, New York 22, N.Y.

GAUTHIER-VILLARS ED.
55 Quai des Grands-Augustins, Paris 6

PERGAMON PRESS G.m.b.H.
Kaiserstrasse 75, Frankfurt am Main

Library of Congress Card Number: 57-59094

Printed in Hungary

CONTENTS

IGC-1959 CALENDAR RECORD

PART III

CALENDAR RECORD FOR THE INTERNATIONAL GEOPHYSICAL COOPERATION 1959

compiled by

A. H. SHAPLEY
CSAGI Reporter for
World Days and Communications
and
J. V. LINCOLN

PREFACE

THIS Calendar Record for the period known as the International Geophysical Cooperation 1959 is a companion to the IGY Calendar Record, published in part I of this volume. It gives in summary form various indices and highlights of the solar, geophysical and associated events which took place during 1959. It thus serves as a sort of expanded index to the data that are available for periods of interest to one or another geophysical discipline. It is not a primary source of data. All of the material collected here has been or will be given in greater detail in other volumes of the *IGY Annals* or in the recognized publications of individual disciplines, or is available in the IGY World Data Centers.

IGC-1959 will likely prove to be as valuable to geophysical science as the IGY. Solar activity remained at a very high level and indeed events occurred which were different from and perhaps even more significant than any occurring during the IGY. Practically all of the IGY synoptic observing stations continued work during the IGC and some stations newly established during IGY produced more reliable data for 1959. Thus, in anticipation that the period of IGC-1959 will also be studied intensively in the coming years and decades, this second Calendar Record has been compiled.

The reader is referred to the Preface to the IGY Calendar Record for the origin of the project to produce a solar and geophysical record in this form. It should be mentioned here that the present and future of this kind of activity is under the immediate auspices of the International Ursigram and World Days Service (IUWDS) in coordination with the Comité Internationale de Géophysique, the latter being the successor to the CSAGI. The IUWDS is an inter-union activity for continuing a World Days program along the general lines of the IGY. IUWDS is the responsibility of the International Scientific Radio Union (URSI) in association with the International Astronomical Union and the International Union of Geodesy and Geophysics. It is also affiliated with the International Council of Scientific Unions' Federation of Astronomical and Geophysical Services. The IUWDS has fostered the completion of this volume and has plans for sponsoring this type of compilation for future periods of geophysical interest.

In making this compilation we have sought and invariably received the cooperation of a great many individuals and institutions all over the world. Indirectly even more have contributed, including many observers at lonely outposts who may not even see their station mentioned explicitly. To mention all is obviously impossible; to mention a few would be impolitic. Our immediate sources for data which we have copied or used in compilations are acknowledged at the appropriate places in the text. In many cases special and time-consuming analyses were made for the purposes of this volume, particularly by scientists of some of the IGY World Data Centers. While every effort has been made to assure that the source data have been copied accurately and interpreted correctly by the compilers of this Record, it is unavoidable that a multi-discipline compilation will have more errors than the separate publications by specialists.

My co-compiler, Miss J. Virginia LINCOLN, is responsible for all the painstaking detail of this Calendar Record and to her goes the credit that the project is finally completed. The huge task of typing the master copy for photo-offset printing of the chart and highlights pages was done by Mrs. Virginia A. NEAR and of the supplementary tables and the manuscript itself by Mrs. Ruth WICHMAN. The considerable amount of drafting and photographic work in preparing the master copy was done at the Central Radio Propagation Laboratory

This Calendar Record for IGC-1959 completes the publications envisaged by CSAGI for the World Days program. It is to be hoped that it, together with the IGY Calendar Record and the Manual for World Days and Communications, have and will aid the joint advance of the IGY disciplines concerned as well as provide a suitable record of this part of the whole IGY and IGC-1959 enterprise.

December 28, 1962.

A. H. SHAPLEY
CSAGI Reporter for World Days and Communications
Central Radio Propagation Laboratory
Boulder, U. S. A.

EXPLANATIONS

I. INTRODUCTION

THE charts and tables in this Calendar Record for the International Geophysical Cooperation 1959 (IGC) give in summary form the solar and geophysical highlights of the period 1 January, 1959 through 31 December, 1959. The data are given in abbreviated form or by indices; explanation, sources and references are given in this text. Both the format and content are similar to that of the IGY Calendar Record, *IGY Annals* XVI pp. 1-157; some exceptions are noted below.

The scientific disciplines involved in the IGC program which are represented in the Record are: Solar Activity, Geomagnetism, Ionosphere, Aurora, Cosmic Rays, Rockets and Satellites.

The Calendar Record itself is made up of pairs of facing pages each pair referring to ten days of the IGC period. The left-hand page is a chart in calendar form and includes pertinent indices and information, and also identification of sunspots crossing the sun's central meridian and a sketch of the location of important solar flares and outbursts. The right-hand page gives day-by-day the highlights of solar and geophysical events taken both from the indices and from other data not suited to systematic treatment; this material includes information on IGC rocket and satellite launchings. Also on the right-hand page is a day-by-day listing of outstanding solar flares.

In the Supplementary Tables are given the IGC Calendar of Regular World Days and Intervals decided on before the IGC began and the declarations of Geophysical Alerts and Special World Intervals which were circulated currently during the IGC as an aid to the conduct of observing programs at stations of many disciplines throughout the world. This material is given for record purposes and as an index to times when geophysical stations may have made intensified or special observations. For future analytical studies of IGC data, The Analysis Intervals in Section VI, below, selected in hindsight, are more appropriate than the Special World Intervals which were selected "on the lun". There are also collected in the Supplementary Tables the ionospheric indices and a list of polar cap absorption events, data which are dispersed throughout the Calendar part of the volume.

The indices given here for IGC 1959 and those given for IGY 1957-1958 in the IGY Calendar Record are homogeneous with the following exceptions:

Under Column 9 of the chart pages (see Section II), "The Fixed Frequency Solar Radio Emission Events" (IGC) are derived from some of the same data sources as the "200 Mc Solar Radio Emission Events" (IGY). However the IGC index takes into account observations at several frequencies and is based on improved criteria which have been developed partly as a result of experience just before and during the IGY. In Column 13, a weighted index of solar flare activity is given for IGC while for IGY only counts of the flares in two ranges of importance were given. In fact the source of definitive information on flares for IGC is different from that used for IGY; this affects Column 12 and the detailed flare list as well as Column 13. For IGY the compilers attempted to interpret the rich but sometimes contradictory basic data in the *IAU Quarterly Bulletin*, while for IGC the McMath-Hulbert collation has been used.

There are a few other differences between the two Calendar Records. The format of Column 6 has been changed and also in a minor way, Column 7. Columns 2 and 3 are different because of changes in the significance of Alers and Special World Intervals. Systematic data on "Spectral Solar Radio Emission Events" (Column 9, top) are added, in addition to the improved indices mentioned above. Dropped are some of the details concerning duplicate observations of solar flares (Section IV, below), references to white light flares and the Supplementary Table giving full details on auroral indices for all sectors and times. Otherwise an attempt has been made to keep the IGC and IGY Calendar Records as much the same as possible, even though some other improved indices are now available, which may eventually be of greater significance than some of those included. Improvements of format have also been deferred to a Calendar Record for some later period.

It must be emphasized that the Calendar Record cannot be considered a primary source of data. Solar and geophysical phenomena are inherently complex and extreme care is necessary not to oversimplify by casual use of indices or abbreviated data. The material in the Calendar Record may suggest interdisciplinary correlations of associations, but deeper research should usually depend on the basic data, such as may be found in the IGY World Data Centers.

II. CHART PAGES

The left-hand of each of the pairs of facing pages which comprise the Calendar Record is a chart covering 10 successive days. The column headings are reproduced in Fig. 1 and the columns are identified by numbers as a key to the description which follows. The source of the data is given in each case and, where appropriate, references to detailed explanations.

1	2	3	4	5	6	7	8	9	10	11	12	13
	Geo-alert	SWI	Magnetic indices Ap C Storm begins ← 3 Hr Kp →	R_z 3000S	Iono-sphere Index	Aur-oral Index	Cos-mic Ray Index	Type IV Out-burst	SWF Imp ≥ 2 <2	Active Region lat. Sunspot lat. Area	Flares Position Imp ≥ 2	Hrs. patrol I_{area}

FIG. 1.

Column 1 (center and top) Calendar date. Regular World Days (RWD) are indicated by a circle around the date, e. g.③(if day of highest priority ⑫). The days in the World Meteorological Intervals (WMI) are indicated by WMI above the date, the Regular World Intervals (RWI) are indicated by the letters RWI above the date; unusual meteoric activity by underlining the date, e. g. *4*; eclipses by a circle and square around the date, e. g. ⑫

Source : International Geophysical Cooperation 1959 Final Calendar; reproduced as Supplementary Table 1 in this volume.

Explanation : IGY Annals, VII, pp. 11-12.

Column 1 (bottom) Identification of magnetically quiet and disturbed days.

Q — one of the 5 quietest days of the calendar month

q — other days which with those marked Q comprise the 10 quietest days of the calendar month

D — one of the 5 most disturbed days of the calendar month

Source — J. VELDKAMP, Koniklijk. Nederlandsch Meteorologisch Instituut, De Bilt, Holland, Committee on Characterization of Magnetic Disturbances, IAGA, IUGG.

Explanation : IGY Annals, IV, p. 235.

Column 2 Geophysical Alert. The significance of Alerts was changed on February 1, 1962, at the end of the period originally envisaged by CSAGI for coordination of observations through the IGY World Days program. During January 1959 the IGY plan was in effect; from February through December 1959, a revised or IGC plan was in effect. The entries in this column are correspondingly different.

(a) *January* 1959: the periods of Solar Alert are indicated by ✓ for 00^h to 16^h UT (to the left) and 16^h to 24^h UT (to the right). In the IGY World Days program, an IGY World Warning Message was issued at 1600UT by the World Warning Agency, and given prompt world-wide distribution. When the level of solar activity seemed to warrant it, the warning message stated "Solar Alert." This warning remained in effect until the next warning message was issued at 1600 UT on the following day at which time the Solar Alert might or might not be renewed. See *IGY Annals*, VII, pp. 12-14, 21-29.

(b) *February—December* 1959: Geophysical Alerts (GEOALERTS) under the IGC World Days program were prompt announcements by the World Warning Agency that an important geophysical (not solar) event had taken place. The entry in this column gives in abbreviated form the category of the Alert and the UT time and date of the beginning of the event; e. g., MAG STORM 11/0756 means a significant magnetic storm started on the 11 th day of the month at 0756 UT. The information on Geophysical Alerts are given here for record purposes; they were, of course, based on preliminary data immediately available to the World Days organization, while definitive data are given under Column 4 and elsewhere in this Calendar Record. However, knowledge of the issuing of GEOALERTS may be useful in planning scientific studies, for many geophysical stations increased their observations when thus alerted and therefore more detailed and special data may be available for such periods.

Source : CRPL-F Report Part B, Nos. 147 through 185; collected in Supplementary Table 2 in this volume.
Explanation : IGY Bulletin (USA), No. 29, May 1959, pp. 12-14.

Column 3 Period of Special World Interval (SWI). The significance of SWI was changed on February 1, 1959, at the end of the period originally envisaged by CSAGI for the coordination of observations through the IGY World Days program. During January 1959 the IGY plan was in effect; from February through December 1959 a revised IGC plan was in effect. The entries in this column are correspondingly different:
(a) *January* 1959: SWI are indicated by ✓ and refer to the full 24 hours of the day. They were declared on a current schedule by the World Warning Agency as periods predicted to be of especial interest to geophysics. The declarations were included in the daily warning message which was given prompt world-wide distribution. See *IGY Annals*, XVI, p. 7, and VII, pp. 12-14, and 21-29.
(b) *February — December* 1959: SWI in the IGC World Days program were declared, as appropriate, at the time of issue of a Geophysical Alert. They therefore designated a period certain to be of sufficient geophysical interest to warrant intensified and special observations. They were declared in the daily warning message issued by the World Warning Agency at 1600 UT. The instruction to start (S) an SWI took effect upon receipt of the warning message at the observing station, but not later than 0001 UT on the day following. The warning message on the next day(s) might declare the SWI to continue (C). Finally, one of the daily warning messages would declare the SWI to be finished (F) at 2359 UT of the date the message was issued. The entries in Column 3 of the Calendar Record chart indicate what the daily warning message at 1600 UT on that day stated concerning SWI — S, C or F.
Source : CRPL-F Report Part B, Nos. 174 through 185; collected in Supplementary Table 3 in this volume.
Explanation : IGY Bulletin (USA), No. 29, May 1959, pp. 12-14.

Column 4 (top) Whole-day magnetic indices and sudden commencements of magnetic storms. Reading from left to right:
Ap index — planetary A-index for the Greenwich day based on a selected standard group of observatories. The index increases with increasing magnetic activity to a maximum of 400.
C index — magnetic character figure C for the Greenwich day based on observations made by a world-wide group of observatories. The index increases with increasing magnetic activity in steps of 0.1 units from 0.0 to 2.0.
ssc — sudden storm commencement with the adopted time in hours and minutes, UT. Immediately below in parentheses is given the number of magnetic observatories which reported the event.
Sources : J. BARTELS (Ap) and J. VELDKAMP (C), Committee on Characterization of Geomagnetic Disturbances, IAGA, IUGG; A. ROMAÑA (ssc), Committee on Rapid Variations and Earth Currents, IAGA, IUGG.
Explanations : IGY Annals, IV, pp. 227-236 and Bulletins of the International Association of Geomagnetism and Aeronomy (IAGA) of IUGG.

Column 4 (bottom) Magnetic Indices Kp. These are the Geomagnetic Planetary Three-Hour Range Indices from a standard group of observatories. The scale increases from 0 to 9 with increasing magnetic activity. Successive indices are given for 00^h-03^h, 03^h-06^h, ... 21^h-24^h UT. The indices are expressed symbolically to 1/3 of a unit; examples: 3- denotes 2 2/3; 3 o denotes 3; and 3+ denotes 3 1/3.
Source : J. BARTELS, Committee on Characterization of Geomagnetic Disturbances, IAGA, IUGG.
Explanation : IGY Annals, IV, pp. 232-236.

Column 5 (top) Relative Sunspot Number, R_z.
Source : M. WALDMEIER, Federal Observatory, Zürich, Switzerland, for the International Astronomical Union.
Explanation : Astron. Mitt. der Eidgen. Sternwarte Zürich No. 215 and 221.

Column 5 (bottom) 3000 Mc S-Index. These solar radio emission flux indices are given in units of watts/M^2/cycle/second bandwidth ($\times 10^{-22}$). They are based mainly on observations by Covington at Ottawa, Canada, and represent closely a continuation of the indices for IGY labelled "Covington 2800 Mc Solar Flux Value" in the IGY Calendar Record, *IGY Annals*, XVI, e.g. p.8.
Source : IAU Quarterly Bulletin on Solar Activity, Nos. 125 through 128.
Explanation : J. Royal Astron. Soc. Canada, 53, pp. 156-161, 1959.

Column 6 (top) Ionospheric World Blackout Index, P.* An index representative of the occurrence of polar-type blackout for each day, UT, which increases approximately as the logarithm of the total number of hours of blackout. The scale is 0 to 9, qualitatively analogous to the magnetic Kp scale. It is based on observations from the higher latitudes in both hemispheres. (Daily indices for each hemisphere separately are given in supplementary Table 4B in this volume.)
Source : W. R. PIGGOTT, Radio Research Station, Slough, U. K.; collected in Supplementary Table 4A in this volume.
Explanation : Some Ionospheric Results Obtained during the International Geophysical Year, pp. 94-101, W. J. G. BEYNON, Editor, Elsevier, 1960.

Column 6 (middle) Ionospheric E-Layer Index, J_E.* This has been derived from data for daylight hours at Slough, U. K., and is mainly representative of the solar ionizing radiation contributing to the production of the E-layer in the period 08^h-16^h UT, daily. Indices enclosed in parentheses, e. g. (238) are doubtful. The letter "C" denotes lack of data, while other missing values indicate that the E-layer was obscured.
Source : C. M. MINNIS, Radio Research Station, Slough, U. K.; collected in Supplementary Table 5 in this volume. (see p. 298).
Explanation : J. Atm. Terr. Phys., 17, pp. 57-64, 1959-1960.

* The order of the indices in Column 6 of this IGC Calendar Record (P, J_E, W) is different from that in the same column of the IGY Calendar Record (J_E, P, W).

Column 6 (bottom) Ionospheric F2-Layer, World Disturbance Index, W.* This is a subjective index on a scale of 0 to 5 of the average degree of disturbance in the F-layer during each day, UT, based on data from 15 to 33 well-spaced stations, and derived according to different qualitative criteria. It is analogous to the magnetic index C. The IGC indices are generally but not strictly comparable to the IGY indices, since fewer data were available for the IGC period.

Source: W. R. PIGGOTT, Radio Research Station, Slough, U. K., collected in Supplementary Table 6 in this volume.

Explanation: Some Ionospheric Results Obtained during the International Geophysical Year, pp. 116-123, W. J. G. BEYNON, Editor, Elsevier, 1960.

Column 7 Auroral Indices. The index is the greatest geomagnetic (gm) colatitude in which aurora appeared overhead in each quarter of the UT day. The index is given for 00-06 UT on the first line, for 06-12 UT on the second line, for 12-18 UT on the third line and for 18-24 UT on the last line. An index was originally prepared for each of eight 45° sectors of longitude — A, B, C, . . . H for the Northern Hemisphere, and a, b, c, . . . h for the southern hemisphere. Only the greatest index for each UT quarter-day is given in this table and the sector (or sectors) to which it applies is identified by the appropriate letter (or letters). Other symbols and conventions used in this column are indicated by the following examples:

31H (index underlined) — this sector (e. g., H) is the only one for which an index is available for this UT quarter-day.

⟨32c (inequality sign) — observational coverage incomplete; overhead aurora in this sector (e. g., c) at gm colatitude less than value given.

Copies of the table of detailed indices, comparable to that appearing in Supplementary Table 4 of the IGY Calendar Record, are available at each of the IGY World Data Centers for Aurora (Visual); see *IGY Annals*, Vol. VII, for locations and addresses of these Centers. The detailed table gives for each sector of longitude, both northern and southern hemisphere, and for each quarter of the UT day, the geomagnetic (gm) colatitude in which aurora appeared overhead.

Source: Auroral World Data Centers and national Auroral Reporters coordinated by S. CHAPMAN, CSAGI Reporter for Aurora and Airglow; C. GARTLEIN (Ithaca), J. PATON (Edinburgh), N. V. PUSHKOV (Moscow), P. MILLMAN (Ottawa), F. JACKA (Melbourne), O. SCHNEIDER (Buenos Aires), A. VAN WIJK (Hermanus), and D. BARBIER (Paris).

Explanation: IGY Annals, XVI, pp. 153-157.

Column 8 Cosmic Ray Indices. Average daily counting rates of four different detectors chosen to be representative of the flux at four points in the cosmic ray spectrum. All of the following cutoff figures are quite rough. "0" denotes instrument not working.

Top number: Data from ionization chamber at Huancayo. The average rigidity of the particles to which this counter responds is about 30 BV.

Second number: Data from neutron monitor at Huancayo, which responds to all particles having a rigidity greater than about 15 BV.

Third number: Data from neutron monitor at Climax, which responds to particles of rigidity greater than 2.5 BV.
Bottom number: Data from neutron monitor at Sulphur Mountain, which has a somewhat lower, but uncertain, cutoff.
Source: Compiled by P. KELLOGG of IGY World Data Center A for Cosmic Rays (Minneapolis) in co-ordination with J. A. SIMPSON, CSAGI Reporter for Cosmic Rays, from data available to the Data Center.
Explanation: IGY Annals, IV, Parts I, II and III, for description of instruments.

Column 9 (top) Spectral Solar Radio Emission Events. One or more Type IV solar spectral events observed at Dapto near Sydney (indicated by Syd), or at Ft. Davis, Texas (indicated by Har).
Source: IAU Quarterly Bulletin on Solar Activity, Nos. 125 through 128.
Explanation: IAU Quarterly Bulletin on Solar Activity.

Column 9 (bottom) Fixed Frequency Solar Radio Emission Events. The index is the number of "radio flares" of importance 2- or greater, compiled by A. D. Fokker from the Netherlands Post Telephone and Telegraph solar radiometers on 200, 545 and 3000 Mc at NERA (Amsterdam), Paramaribo and Hollandia.
Source: A. D. FOKKER, Utrecht University, Netherlands.
Explanation: Bulletin of the Astronomical Institute of the Netherlands XV, No. 507, pp. 319-324.

Column 10 Short-Wave Fadeouts (SWF). This gives the number of independent SWF observed each day from all available reports, edited by J. V. LINCOLN. No distinction is made between sudden, gradual and other types of SWF. Most of the data are from signal intensity recordings of distant high frequency radio transmissions. There is no assurance of 24-hour or even uniform coverage of these phenomena. The SWF are classed subjectively into categories of importance. The top line gives the number of SWF of importance greater than or equal to class 2; the bottom line gives the number of SWF of importance less than class 2.
Source: CRPL Compilations of Solar-Geophysical Data, available from IGY World Data Center A for Ionosphere (Boulder).
Explanation: National Bureau of Standards Report 5540.

Column 11 (upper part) Date of Passage of Active Solar Regions across Central Meridian. Regions designated as Active Regions in the *IAU Quarterly Bulletin on Solar Activity* are identified by their latitude (e. g. S15 means 15° south heliographic latitude) printed in plain type on the day of passage of the sun's Central Meridian (CMP). When two or more active regions have CMP on the same day, they are listed in the order of passage, according to their nominal central positions. The definition of active region for this purpose is generally that the region is the seat of solar flares; the size of the sunspots in the region is not a criterion.
Source and Explanation: IAU Quarterly Bulletin on Solar Activity.

Column 11 (lower part) CMP of Large Sunspot Regions and their Areas. Sunspot regions with areas greater than 500 millionths of a solar hemisphere* are identified by their latitude entered in slant *(italic)* type on the day of passage of the sun's Central Meridian (CMP). Immediately below is given the area of the region at the observation nearest to CMP, expressed in millionths of a solar hemisphere. These large sunspot regions are not necessarily the seat of flare activity.
Source : Data provided by Royal Greenwich Observatory.
Explanation : Greenwich Publications.

Column 12 Sketch of Sun Showing Positions of Flares. The circle represents the sun, with heliographic North, South, East and West as indicated. Shown in their approximate positions on the disk are the H-alpha flares of importance ≥ 2, with additional symbols to indicate whether certain other phenomena were observed in time association with the flare, namely either a short-wave fadeout (SWF), or a fixed frequency solar radio emission event (burst) of importance 2- or greater (SRB), or both. The symbols have the following meaning:

● Flare of importance ≥ 2; no SWF or SRB reported
⊙ Flare and SWF; no SRB reported
× Flare and SRB; no SWF reported
⊗ Flare and both SWF and SRB

Source : Flares are from McMath-Hulbert Observatory Working List of Flares for IGC 1959 (see Column 13). SWF and SRB data are same as for Columns (10) and (9, bottom), respectively. Compilation by J. V. LINCOLN.

Column 13 (top) Completeness of Solar Observations. The index is the number of hours during the 24 hour day during which the sun was under observation by a systematic flare patrol at one or more observatories. This index gives an indication of the likelihood that an outstanding solar flare was missed.
Source : IAU Quarterly Bulletin on Solar Activity, Nos. 125-128, collated by J. V. LINCOLN.

Column 13 (bottom) Flare Index. The number is the daily flare index, I_{Area}, produced by H. W. Dodson and E. R. Hedeman from the data in the McMath-Hulbert Observatory Working List of Flares for IGC-1959. The index is the sum for the UT day of N_i n_i, where N_i is the number of flares of importance i, and w_i is the weight assigned to that importance category. The weights, w_i, are approximately proportional to the average area of flares in the respective importance category, namely 1, 2.5, 5 and 10 for flare importance 1, 2, 3 and 3+ respectively. These flare indices were not given for IGY in the IGY Calendar Record, but IGY indices are listed in IGY Solar Activity Report Series (World Data Center A), Number 14.

* 1 square degree (heliographic) has an area of 48.5 millionths of the solar hemisphere.

Source and Explanation : McMath-Hulbert Observatory Working List of Flares and Daily Flare Index for IGC-1959, IGY Solar Activity Report Series No. 15, World Data Center A, Boulder, 1961.

III. ARRANGEMENT OF HIGHLIGHTS FOR EACH DAY

On the page facing the chart for each ten days, is an abbreviated, but mainly self-explanatory text giving some of the highlights of solar and geophysical events which took place on each of those days, and other information of interest, not suitable for presentation on the chart page. The criteria for inclusion in this section have been set by the compilers.

The highlights always appear in the following order. If a category is omitted for any day, this means nothing of note occurred or was reported.

(1) Level of magnetic activity
(2) Outstanding aurora
(3) Cosmic ray events
(4) World blackout days
(5) Polar cap absorption events
(6) Selected solar phenomena — surges, dark filaments, prominences
(7) Spectral radio emission
(8) Dates of observation of solar electron corona
(9) IGC rocket launchings
(10) IGC satellite and space probe launchings

(1) *Level of Magnetic Activity.* A descriptive statement of the degree of magnetic activity. If the day is one of the 5 or 10 quietest days of the month or one of the 5 most disturbed days of the month, this is so stated in order to call attention to the designation in Column 1 of the facing chart page. If the Ap index for the day exceeded 100, the day is identified as one of the most disturbed days of the IGC. There are eight such days. Nineteen days are identified as the quietest days of the IGC. For this, all of the following criteria had to be satisfied: the Ap index for the day less than 5, the C index less than 0.2, and the day not adjacent to a disturbed period.

(2) *Outstanding Aurora.* The words "outstanding aurora" appear for days when one or more of the quarter-day indices in Column 7 was greater or equal to 35 but less than 40 in northern hemisphere sectors A, C, D, E, G, H and in all southern hemisphere sectors, or greater or equal to 29 but less than 34 in northern hemisphere sector B. The criteria for the statement "very outstanding aurora" are ≥ 40 and ≥ 34, respectively.

(3) *Cosmic Ray Events.* The words "Forbush cosmic ray decrease" appear on the days for events given in *Technical Note* No. 5. *Uppsala University, Institute of Physics, Cosmic Ray Group.* The sea level increase in cosmic rays July 17 was described in *Space Science Reviews*, Vol. 1, p. 31, 1962.

(4) *World Blackout Days.* These are the five most disturbed days each month according to the ionospheric world blackout index (P) in Column 6 (top).

(5) *Polar Cap Absorption Events.* The approximate time of beginning, duration and magnitude of the event are taken from a list by C. S. WARWICK and M. W. HAURWITZ *(J. Geophys. Res.* 67, p. 1322, 1962). These data are also collected in Supplementary Table 7 in this volume.

(6) *Selected Solar Optical Phenomena.* Lists of outstanding solar events (other than flares, which are clearly indicated elsewhere on these pages) have been provided by R. HANSEN, IGY World Data Center A for Solar Activity, selected by him from the data in *IGY Solar Activity Report Series,* Numbers 8, 9, 10, 11 of IGY WDC-A for Solar Activity. The category of event is specified, its nominal location in terms of heliographic coordinates or position angle, and the nominal time it took place. The categories of phenomena are dark surge on disk, bright surge at limb, active prominence region, active filament region, active dark filament, bright surge on disk and sudden disappearance of filaments.

(7) *Spectral Radio Emission.* Radio spectra observations of the sun provide details of many events of potential interest. A selection has been made by J. V. LINCOLN from data published in the *IAU Quarterly Bulletin on Solar Activity* for Ft. Davis, U. S. A. (range 100-580 Mc) and Sydney, Australia (nominal range 40-200 Mc). The times of Type II and Type IV bursts are indicated.

(8) *Dates of Observation of Solar Electron Corona.* This calls attention to the existence of observations of the solar electron corona at Climax, U. S. A. by High Altitude Observatory, usually between 1700 and 2100 UT. This entry does not refer to a solar or geophysical event; it does identify the only 56 dates during the IGC when atmospheric and other conditions made successful observations possible. These observations are described in *IGY Solar Activity Report Series* Number 16 of IGY WDC-A for Solar Activity.

(9) *IGC Rocket Launchings.* Upper atmosphere sounding rocket experiments are identified by country, rocket number or name, type of experiment, altitude reached (if information available), time of launch, and launching site. The material for these entries was obtained from IGY World Data Center A, Rockets and Satellites *Catalogue of data received by WDC-A during the period 1 July 1957—31 December 1961.*

(10) *IGC Satellite and Solar Probe Launchings.* The dates of IGC satellite and solar probe launchings are taken from IGY World Data Center A, Rockets and Satellites *Catalogue of data received by WDC-A during the period 1 July 1957—31 December 1961,* and *COSPAR Information Bulletin* No. 2. For satellites the Greek letter designation, popular name, type of experiment, and estimated life time are given. For space probes the popular name, type of experiment and trajectory are given.

IV. DETAILED LIST OF SOLAR FLARES

At the bottom of the pages of "Highlights" is a day-by-day listing of important flares observed. The criterion for inclusion is that the flare must have been rated in importance class 2 or greater in the *McMath-Hulbert Observatory Working List of Flares for IGC 1959,* (see II (13)) which is also the source of the detailed data given. Listed are (1) the importance rating assigned; (2) the beginning and ending times in UT; (3) the nominal heliographic co-ordinates of the flare and (4) the total number of observatories which reported the event. Convention and symbols are as follows:

(*a*) Beginning and ending times are hours and minutes in UT, although in the table UT is omitted. The letter "E" (e. g. 0714E) means "less than" or "before" the time given; the letter "D" (e. g. 1315D) means "greater than" or "after" the time given.

(*b*) Heliographic co-ordinates are expressed in degrees north and south of the sun's equator (e. g. S13) and in degrees east and west of the sun's central meridian (e. g. W62).

V. DESCRIPTION OF SUPPLEMENTARY TABLES

It has been convenient to repeat in the Supplementary Tables material in compact form which is spread out in the Calendar Record.

Supplementary Table 1 — IGC Calendar. This is a reprint of the Final IGC Calendar widely distributed during the period of IGC observations.

Supplementary Table 2 — IGC Geophysical Alerts. This is a complete list of the periods of solar and geophysical alert given individually in Column 2 of the chart pages of the Calendar Record. Further information and references appear in II(2) above.

Supplementary Table 3 — IGC Special World Intervals. This is a complete list of the Special World Intervals given individually in Column 3 of the chart pages of the Calendar Record. Further information and references appear in II(3) above.

Supplementary Table 4 — Blackout Indices. The World Blackout Index (P) which is given for each day in Column 6 (top) on the chart pages of the Calendar Record is retabulated compactly in Table 4A, together with identification of the five most disturbed days in each month according to this index. Separate indices for north and south hemispheres for each day are given in Table 4B. Further information and references appear in II(6) above.

Supplementary Table 5 — E-Layer Indices. In this table the E-layer Index J_E which is given for each day in Column 6 (middle) on the chart pages of the Calendar Record, is retabulated compactly. Further information and references appear in II(6) above.

Supplementary Table 6 — F2 Disturbance Index. In this table the F2 Disturbance Index (W) which is given for each day in Column 6 (bottom) on the chart pages of the Calendar Record, is retabulated compactly. Further information and references appear in II(6) above.

Supplementary Table 7 — Polar Cap Absorption Events. Collected here is the complete list of Polar Cap Absorption Events which are listed individually on the highlights pages of the Calendar Record. Further information and references appear in III(5) above.

VI. ANALYSIS INTERVALS

The material of the Calendar Record has been examined in some detail, and selections have been made of IGC Analysis Intervals, or periods of time for which detailed and interdisciplinary analysis of observation would seem to be especially fruitful. Such intervals have been selected by the CSAGI Reporter for World Days and appear in Table 1 below. The selections include periods of relatively quiet conditions as well as disturbed periods.

The criteria for selecting disturbed periods were (1) magnetic storms with high Ap; (2) occurrence of Forbush cosmic ray decreases; (3) occurrence of polar-cap absorption events. The 6 disturbed IGC Analysis Intervals include the 5 magnetic storms of the IGC with greatest Ap daily index, ranging from 103 to 236 and an interval February 25-26 chosen by the U. R. S. I. World-wide Soundings Committee. They also include all but one of the Forbush cosmic ray decreases and all of the polar cap absorption events.

The 14 quiet IGC Analysis Intervals have been selected by the following criteria: (1) Ap index less than 5 and C index less than 0.2, at the same time; (2) isolation in time from a severe disturbance. They include a quiet interval February 23-24 chosen by the U. R. S. I. Worldwide Soundings Committee.

TABLE 1: ANALYSIS INTERVALS–1959

Disturbed		Quiet	
*Feb.	25–26	Jan.	1–2
Mar.	26–29	Jan.	20–21
May	10–12	Jan.	24
July	10–19	*Feb.	23–24
Aug.	16–17	Mar.	9–10
Sept.	2–4	Mar.	16
		Apr.	22
		May	28–29
		June	13
		July	30
		Aug.	28
		Oct.	28
		Nov.	20
		Dec.	21

* Chosen by Worldwide Soundings Committee of U. R. S. I.

CALENDAR RECORD
CHART PAGES AND HIGHLIGHTS

1959

JANUARY	GEO-ALERT	SWI	MAGNETIC INDICES Ap C STORM BEGINS / ← 3 HR Kp →	R_Z / 3000S	IONO-SPHERE INDEX	AUR-ORAL INDEX	COS-MIC RAY INDEX	TYPE IV OUT-BURST	SWF IMP ≥2 <2	ACTIVE REGION LAT. SUNSPOT LAT. AREA	FLARES POSITION IMP ≥2	HRS. PATROL / I_{AREA}
1 Q	✓✓		1 0.0 1122 (38) 0o0o0+1- 1-0+0o0o	221 225	1 (312) 1	- 28E 31E 30E	506 506 504 499	1	1	N06	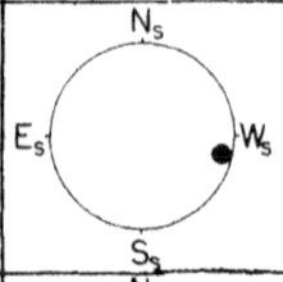	20 6.5
2 Q	✓✓		3 0.1 0o0o0+2- 2-1+1o1o	225 240	0 335 0	- 21G 30E 28E	506 506 506 503		1 1		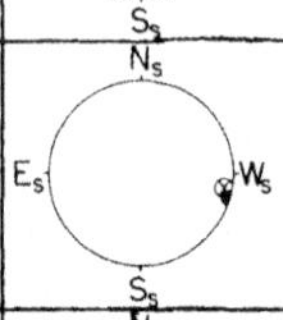	22 14
(3) q	✓✓		6 0.2 2o2+2+2o 2o1o2-0+	229 264	0 308 0	32H 29E 30D/E 32D	505 506 506 502		2	S17 N04	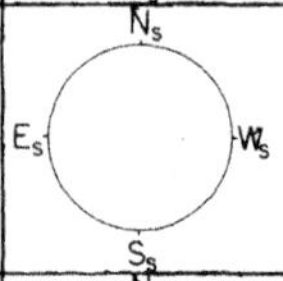	22 7
(4) q			7 0.5 1o1-2-2o 2+3-2o1+	231 274	1 349 1	31D 29E/G 36D 36D	505 506 506 501	1	1	S15 N21	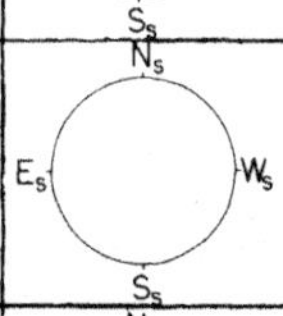	18 3
5			25 1.3 0136 (65) 1029 (2) 2o1+1o4- 4o4o5-6-	243 285	2 (325) 2	31D 36E 39D 40E	503 503 505 499	1	1	N15	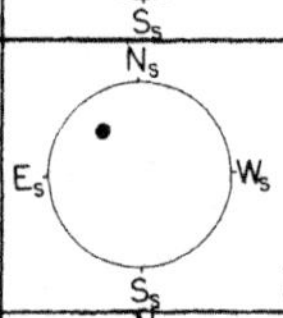	23 10.5
6 D			24 1.2 5-4o3o3+ 4o4-4o3+	238 291	4 327 3	38D 31D/E 38D 38D	502 503 501 489			S16	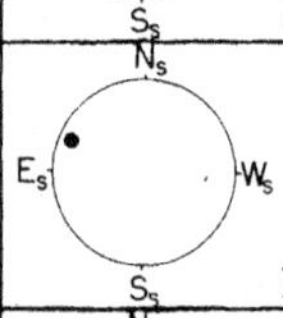	23 10
7 D			19 1.1 4+4-3+3o 2-2o4-4o	247 212	2 270 3	33E 31D 33D 36C	501 503 501 492	Syd		S11 N16 N22 S08 N17 1575	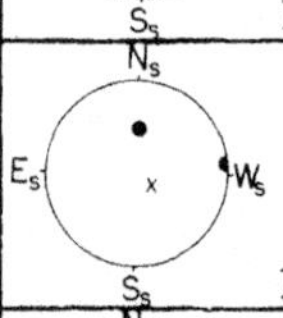	20 10.5
8			18 1.1 4o3o2+3o 2+3o4-4+	246 274	1 298 2	34D 47d/e 35C 37D	501 503 502 502			N18	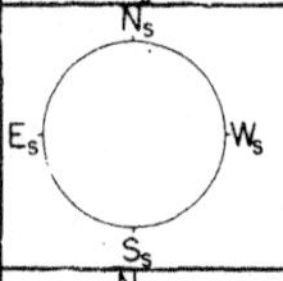	24 1
(9) D			38 1.5 1459 (48) 5-2+2o3- 4-6o5-6+	245 270	3 288 3	33D/E 28E 41D 41D	500 499 500 497	1		S15 N17 S04	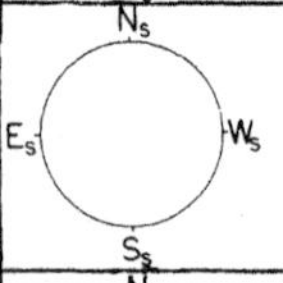	24 7
(10) D	✓✓		45 1.6 6-5o5-5o 5-4+5+4+	224 284	4 260 5	38D 35D 39E 42E	500 498 489 482			S26 N12 3075	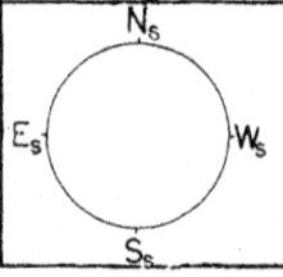	24 7

Jan. 1 One of five quiet magnetic days - one of quietest days in IGC. Bright surge limb, 1556-1626 UT, 18S 90W.

Jan. 2 One of five quiet magnetic days - one of quietest days in IGC. Bright surge limb, 1425-1510 UT, 16S 85W. U.S.S.R. launched space probe Lunik I; corpuscular radiation, magnetic fields, micrometeorites, interplanetary gas; passed within 750 km of moon, now in solar orbit with 15-month period.

Jan. 3 One of ten quiet magnetic days. Spectral radio emission II, 1610.3-1613 UT.

Jan. 4 One of ten quiet magnetic days. Outstanding aurora. Bright surge limb, 0322-0420 and 1557-1650 UT, 15S 90W.

Jan. 5 Moderate magnetic activity with sudden commencement at 0136 UT. Very outstanding aurora.

Jan. 6 Moderate magnetic activity - one of five disturbed days. Outstanding aurora. One of five disturbed World Blackout days. Dark surge disk, 0029-0041 and 0136-0218 UT, 20N 70E.

Jan. 7 Slight magnetic activity - one of five disturbed days. Outstanding aurora. Spectral radio emission II, 0245-0247 UT; IV, 0246-0443 UT.

Jan. 8 Slight magnetic activity. Very outstanding aurora. Sudden disappearance of filament, 2100-2320 UT, 30S 90E.

Jan. 9 Moderate-to-severe magnetic storm with sudden commencement 1459 UT - one of five disturbed days. Very outstanding aurora. One of five disturbed World Blackout days.

Jan.10 Moderate magnetic activity - one of five disturbed days. Very outstanding aurora. One of five disturbed World Blackout days.

Hα flares

Jan. 1 - 2, 0850-0950, S13 W62, 7

Jan. 2 - 2, 0911-1002, S10 W69, 7; 2, 1421-1507, S15 W82, 2

Jan. 5 - 2, 1049-1210, N26 E27, 7

Jan. 6 - 2, 0157-0204, N17 E56, 1

Jan. 7 - 2, 0215-0404D, S10 W10, 2; 2, 0637-0725, N07 W87, 2;
2, 1750E-2045, N25 W03, 2

1959

JANUARY	GEO-ALERT	SWI	MAGNETIC INDICES Ap, C	STORM BEGINS	← 3 HR Kp →	Rz / 3000S	IONOSPHERE INDEX	AURORAL INDEX	COSMIC RAY INDEX	TYPE IV OUTBURST	SWF IMP ≥2 <2	ACTIVE REGION LAT. SUNSPOT LAT. AREA	FLARES POSITION IMP ≥2	HRS. PATROL / I AREA
11	✓✓		14 1.0		5-2+2o1+ 2+2o3o4-	218 251	2 287 4	33D 34E 34E 39E	502 501 495 488		1	N12		19 5
12	✓✓		13 0.9	1124 (8)	4o2+1o2o 3-3o3o3o	203 249	1 288 2	34D 29E 33D 37D	503 504 500 495		1	N23		22 6
13	✓✓		9 0.5		2-2o3-3- 2-2+3o1o	192 234	1 313 1	28C 33G 32E 33D	505 505 505 497			N25		22 15
14 q	✓✓		8 0.5		0+1-3o2+ 2+1+2-3-	128 220	0 (276) 1	28D 30^{E}_{G} 33D 34E	505 503 503 504		2			23 10
15			8 0.3		2o4-3o2+ 1+1o1-1-	120 204	0 315 1	33D 33E 32f 33D	505 506 508 506		1	N23 S04		21 6
16			17 1.2	0624 (2) 0927 (42)	1+0+2o4+ 4o4-4-4-	143 210	3 316 2	34H 33^{D}_{E} 40D 36D	505 504 506 502					22 7
17			18 1.0		3-4o3+3+ 4o4-3-2o	168 214	2 (285) 3	29G 32H 33^{D}_{E} 33E	505 505 508 507					24 10.5
18			19 1.0		4-4o3o3- 3o4o3o3+	179 225	3 2	31H 31f 33D 34D	504 503 510 504	1				22 2
19			10 0.4		2o3o3-3- 2+2o3-1+	202 246	2 (287) 1	28^{C}_{D} 28^{E}_{G}H 33E 31D	504 504 512 507		4			24 13
20 Q			4 0.1		1-0o1o1+ 1o1o1+2-	240 284	1 291 1	27C 34E 27D 31E	505 505 514 510			N12		24 8

Jan.11 Slight-to-moderate magnetic activity. Outstanding aurora.

Jan.12 Slight magnetic activity with sudden commencement 1124 UT. Outstanding aurora. Dark surge disk, 2100-2155 UT, 309° position angle, 0.67 radius vector.

Jan.13 Slight magnetic activity.

Jan.14 Slight magnetic activity - one of ten quiet days.

Jan.15 Slight magnetic activity.

Jan.16 Slight-to-moderate magnetic activity with sudden commencements 0624 UT and 0927 UT. Very outstanding aurora. One of five disturbed World Blackout days.

Jan.17 Slight-to-moderate magnetic activity.

Jan.18 Slight-to-moderate magnetic activity. One of five disturbed World Blackout days.

Jan.19 Slight magnetic activity.

Jan.20 One of five quiet magnetic days - one of quietest days in IGC.

Hα flares

Jan.14 - 2, 0225-0456, S24 W44, 2; 2+, 1406-1516, N28 W12, 1

Jan.17 - 2, 0348-0416, S08 E68, 1

Jan.19 - 2, 0010-0122, N18 E34, 2; 2, 0230-0250, S15 E50, 1

1959

JANUARY	GEO-ALERT	SWI	MAGNETIC INDICES Ap	C	STORM BEGINS	← 3 HR K_p →	R_Z / 3000S	IONO-SPHERE INDEX	AUR-ORAL INDEX	COS-MIC RAY INDEX	TYPE IV OUT-BURST	SWF IMP ≥2 / <2	ACTIVE REGION LAT. / SUNSPOT LAT. AREA	FLARES POSITION IMP ≥2	HRS. PATROL / I_{AREA}
21 Q			3	0.1		0+0o1o1o 1+1o1+0+	255 298	0 374 1	25D 23G 30 D/E 32D	506 507 512 509	1		N17 N12 N18 *N18 1475*	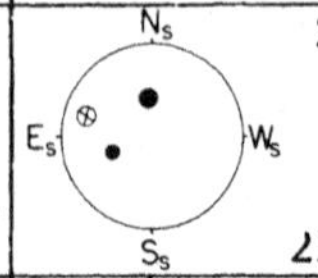	22 23
22 q	✓✓		7	0.4		1-2o3o2- 2-2-2-2-	278 320	0 (319) 0	- 28 D/E 29 D/E 29D	505 506 511 506	3		N00 S10 N10 *N08 1950*	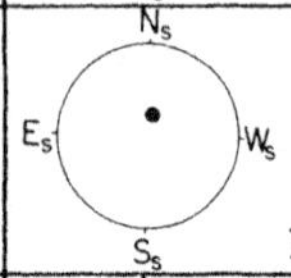	21 15.5
23 q	✓✓		8	0.5		3o2-1+3- 3o1+1+1o	270 274	0 (288) 0	20A 30E 26D 29D	506 508 510 510			N20 S12	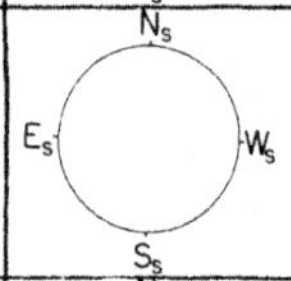	24 9
24 Q	✓✓	✓	4	0.1		2-2-2o1- 1-1o0+0+	261 292	0 - 0	<30c - 33E 33E	507 509 509 510		2		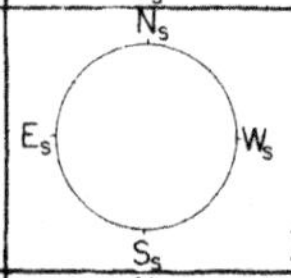	23 12
25	✓✓	✓	18	1.3	0859 (60) 1124 (5)	1o1o1-4- 4+4+4+3-	255 303	1 - 2	- 26G 37D 36D	507 509 510 507		2 1	N28 N13 S07 *N12 1600*	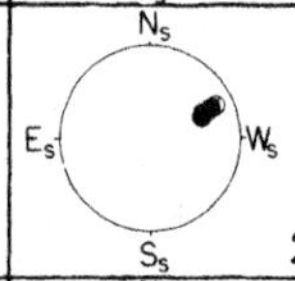	22 21.5
26 D	✓✓		22	1.2	0322 (25)	1+4-2-4+ 6o4-1+1o	252 299	2 - 3	23G 30f 36D 31E	503 504 505 502		1	N21	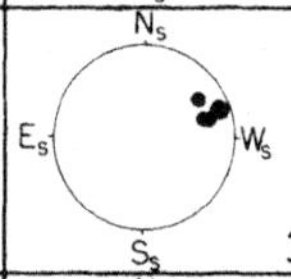	23 38.5
27	✓✓		9	0.6	0914 (4)	0o2o3-3o 3-2+2o2o	263 312	2 (397) 2	- 29E 31E 35E	496 496 491 489	2	1 1	N22	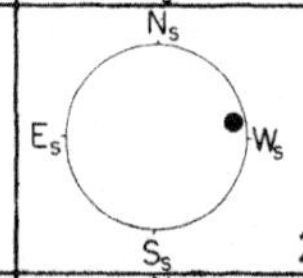	23 20.5
28	✓✓		9	0.5		3o2+2-2o 2o2+3-2+	239 272	0 365 1	27H 28E 33E 31D	497 496 491 485		2	S09	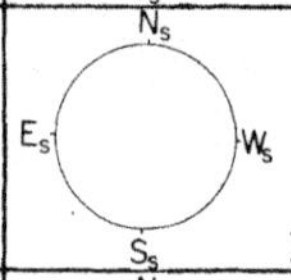	24 17
29	✓✓		17	1.2	0704 (3)	1+2-3-3+ 3o3+4+4o	213 242	1 320 2	33E 35f 34E 34 C/E	497 494 485 483				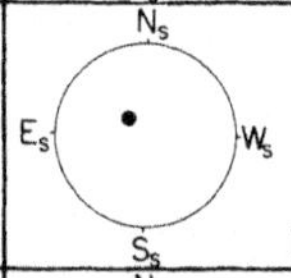	23 10.5
30			11	0.7		3+3+2+3o 2+2o2+0+	167 224	1 (304) 2	36D 33E 34D 35D	499 499 493 492			N03 N05	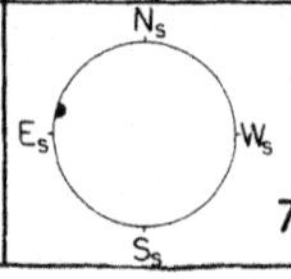	23 7.5

Jan.21 One of five quiet magnetic days - one of quietest days in IGC. Spectral radio emission II, 1717.6-1743 UT.

Jan.22 Slight magnetic activity - one of ten quiet days. Spectral radio emission II, 2101.9-2112 UT.

Jan.23 Slight magnetic activity - one of ten quiet days.

Jan.24 One of five quiet magnetic days - one of quietest days in IGC.

Jan.25 Slight-to-moderate magnetic activity with sudden commencements 0859 UT and 1124 UT. Outstanding aurora.

Jan.26 Slight-to-moderate magnetic activity with resurgence at 0322 UT one of five disturbed days. Outstanding aurora. Active prominence region, 1740-1820 UT, 080° position angle.

Jan.27 Slight magnetic activity with small resurgence at 0914 UT. Outstanding aurora.

Jan.28 Slight magnetic activity. Active prominence region, 0910-1020 UT, 20N 90W.

Jan.29 Slight-to-moderate magnetic activity with sudden commencement at 0704 UT. Outstanding aurora.

Jan.30 Slight magnetic activity. Outstanding aurora. Bright surge limb (or flare), 2342-2410 UT, 076° position angle.

Hα flares

Jan.21 - 2, 0330-0434D, N20 E04, 3; 2, 1246-1436D, S13 E27, 4; 3, 1700-1750, N10 E48, 2

Jan.22 - 2, 1108-1147, N08 W03, 7

Jan.25 - 2, 0210-0315, N16 W45, 5; 2, 0910E-0950, N17 W48, 1; 2, 1406-1450, N18 W50, 6; 2, 1423-1455, N13 W36, 4; 2, 1925-2045, N10 W36, 2

Jan.26 - 2, 0013E-0110, N09 W42, 1; 2, 0647E-0735, N21 W40, 1; 2, 0647E-0810, N08 W47, 2; 3, 0842-1000, N16 W61, 9; 3, 1027-1315, N16 W61, 8

Jan.27 - 2+, 1340E-1520, N08 W60, 11

Jan.29 - 2, 1439-1503, N04 E12, 6

Jan.30 - 2, 2341-0010, N13 E90, 2

1959

JANUARY	GEO-ALERT	SWI	MAGNETIC INDICES Ap	C	STORM BEGINS	← 3 HR K_p →	R_z / 3000S	IONO-SPHERE INDEX	AUR-ORAL INDEX	COS-MIC RAY INDEX	TYPE IV OUT-BURST	SWF IMP ≥2 / <2	ACTIVE REGION LAT. / SUNSPOT LAT. AREA	FLARES POSITION IMP ≥2	HRS. PATROL / I_{AREA}
31			14	0.8		3o4o4-3+ 2o2-2o2-	143 213	2 320 2	29^{G}_{H} 30^{DG}_{EH} 32D 31D	501 502 498 500	1	1	N21 N15	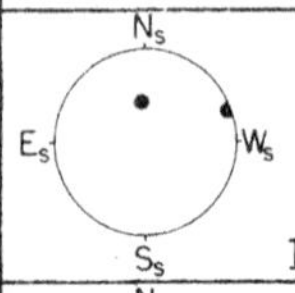	21 11
FEB 1 q			11	0.7		3-3o2+3- 3+3-2-1+	141 195	2 327 1	29D 33E 34D 34D	502 503 501 503	1	1 1	S12 N17	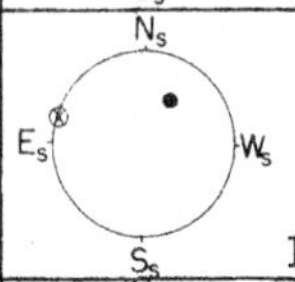	19 13.5
2			23	1.1		2-4+4+4o 4o3+3+4o	141 213	2 343 2	31^{C}_{H} 34E 40D 37D	503 505 508 505	1	1 1		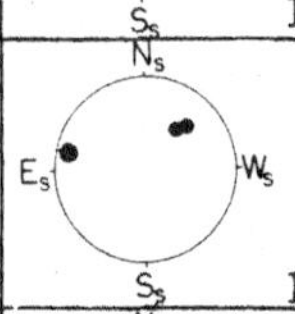	23 18.5
3			22	1.2		4o3-2+4o 4-4-3+5-	140 202	1 316 3	34D 33^{D}_{E} 36^{D}_{E} 35^{D}_{E}	502 504 515 503		1	N22 N14 N20	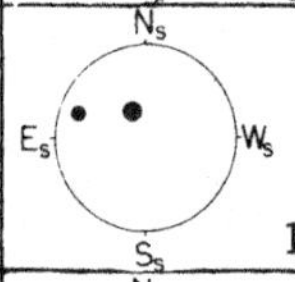	18 13
4 D			36	1.3	0550 (2)	5o4+3+4o 5-4o5-5+	137 213	3 351 2	35D 32E 36E 39D	501 501 513 499			N23 N21 N10	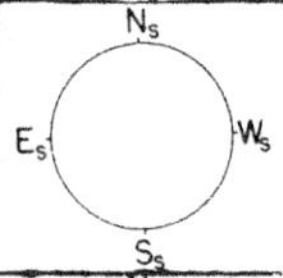	23 3
5			24	1.1		4-5-4-4- 4-4o3+4-	128 191	2 285 3	35D 32^{E}_{G} 38D 35E	499 499 508 496			S28	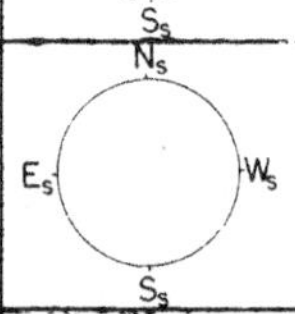	23 5
6			18	1.0		4o3-2o4- 3+3+4o3o	114 200	2 289 3	34^{C}_{D} 29G 38C 34^{D}_{E}	500 499 507 496				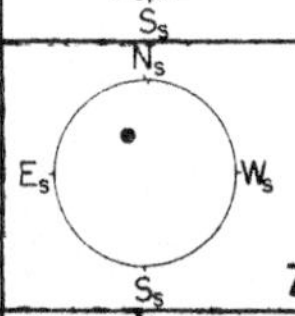	23 7.5
7 q			11	0.6		3o3-2-2+ 3+2+3-1o	124 194	1 272 2	34D 29^{E}_{H} 33D 34C	502 502 510 502			N13	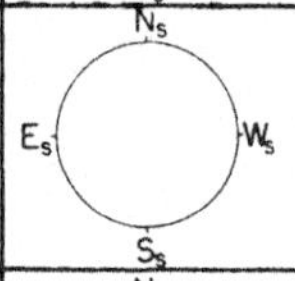	20 6
8 q			12	0.9		1+2-2+3- 3o3-3-4o	90 184	1 C 0	34D 33D 39f 33^{C}_{DE}	503 502 511 502	1	2 1	N08	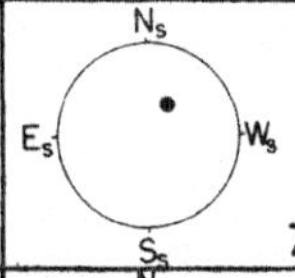	13 7.5
9			23	1.1		5-5-5o3+ 3+3-1o2o	87 189	2 303 1	33^{C}_{D} 33G 35E 34D	503 504 512 497	Syd 4	3 2		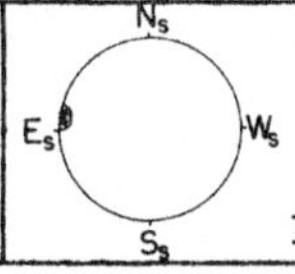	22 13.5

Jan.31 Slight-to-moderate magnetic activity.

Feb. 1 Slight magnetic activity - one of ten quiet days. Bright surges at at limb, 0426-0437, 0443-0511 and 0546-0607 UT, 17N 90E.

Feb. 2 Slight-to-moderate magnetic activity. Very outstanding aurora. Spectral radio emission II, 1853.1-1903 UT.

Feb. 3 Moderate magnetic activity. Outstanding aurora. Spectral radio emission II, 0054-0106 UT.

Feb. 4 Moderate magnetic activity with sudden commencement 0550 UT - one of five disturbed days. Outstanding aurora. Active filament region, 0319-0526 UT, 09N 02W.

Feb. 5 Moderate magnetic activity. Outstanding aurora. Sudden disappearance of filament, 2145-2240 UT, 11N 90W.

Feb. 6 Slight-to-moderate magnetic activity. Outstanding aurora.

Feb. 7 Slight magnetic activity - one of ten quiet days. Spectral radio emission II, 2334.3-2341 UT.

Feb. 8 Slight-to-moderate magnetic activity - one of ten quiet days. Outstanding aurora.

Feb. 9 Moderate magnetic activity. Outstanding aurora. Active prominence region, 0824-1329 UT, 10N 90E. Spectral radio emission II, 0208-0245 UT; IV, 0215-0317 UT.

Hα flares

Jan.31 - 2, 0108-0152, N20 E03, 4; 2, 0840-0945, N19 W79, 2

Feb. 1 - 3, 0352-0458, N12 E83, 1; 2, 1153-1233, N23 W17, 2

Feb. 2 - 2, 0232-0315D, N20 W22, 1; 3, 1017-1225, N22 W30, 7;
3, 1816-2055, N09 E60, 3

Feb. 3 - 2, 0040-0107, N12 E08, 1; 2, 1132-1215, N13 E50, 7

Feb. 6 - 2, 1442-1602, N18 E11, 3

Feb. 8 - 2, 0309-0340D, N14 W13, 1

Feb. 9 - 2, 0058E-0246, N10 E90, 1; 2+, 0943-1019, N09 E88, 10;
2, 1230-1440D, N09 E88, 5

1959

FEBRUARY	GEO-ALERT	SWI	MAGNETIC INDICES Ap C STORM BEGINS ← 3 HR K_p →	R_Z 3000S	IONO-SPHERE INDEX	AUR-ORAL INDEX	COS-MIC RAY INDEX	TYPE IV OUT-BURST	SWF IMP ≥2 <2	ACTIVE REGION LAT. SUNSPOT LAT AREA	FLARES POSITION IMP ≥2	HRS. PATROL I_{AREA}
10 Q			6 0.2 2+1o0+1o 1-1+1+3-	100 190	1 288 1	33^{C}_{H} 33G 31D 33D	505 509 516 509		1 2	N24 N30	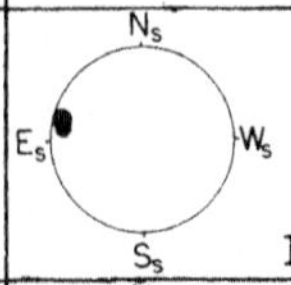	23 18
11	MAG STORM 11/ 0756	S	36 1.4 0318 (66) 0756 (55) 1o4-4-6+ 5+5-4-4+	100 199	3 295 2	30H 36E 38E 38C	502 480 509 501			N20 N32	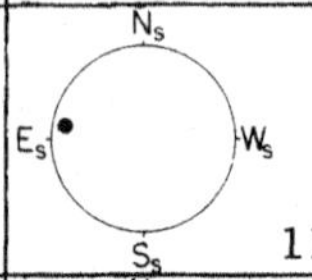	21 11.5
12		F	24 1.1 4-5o3+4o 3+3+3o4-	101 206	3 264 3	32H 33G 37E 34^{D}_{E}	493 490 489 484	1	1		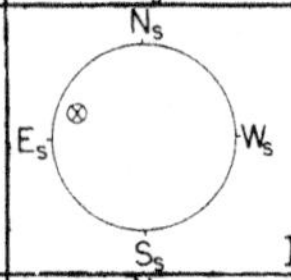	22 18
13			21 1.2 4-2o1+3+ 4+4o4+4o	106 210	4 261 2	32D 29^{E}_{G} 36E 38D	494 490 486 482	Har		N25	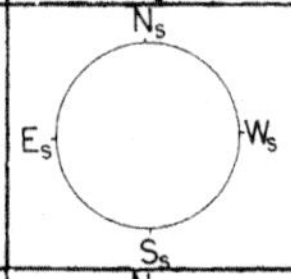	19 6
14			30 1.3 1142 (51) 3+4o4-4o 6-5o3+2+	129 210	6 233 3	34D 32E 37^{D}_{E} 38^{C}_{D}	493 489 481 477			N20	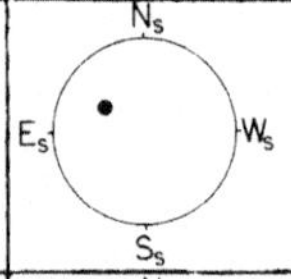	23 7.5
15			37 1.3 2350 (2) 3+6-6-5o 4+4+4o2-	133 217	7 243 5	32^{D}_{H} 34H 37E 38D	490 484 464 459		2	*N08* *580*	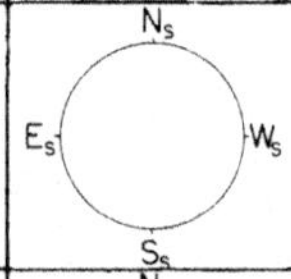	17 7
16 D			61 1.7 0649 (15) 5o6o5o3o 5+6-6+6o	144 222	5 251 5	33D 33H 36^{D}_{E} 51D	492 486 466 461		2	N09 S35	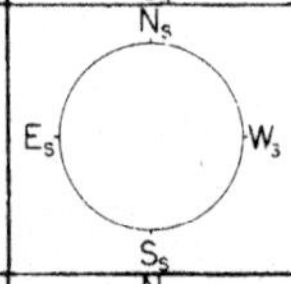	22 5
⑰			24 1.0 6-5-5o3- 2o2o2o1-	170 233	5 225 4	38C 33H 30E 38D	493 487 466 462		1	N17	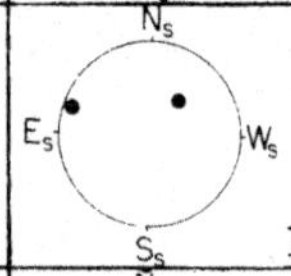	23 15
⑱ Q			4 0.1 1+1-1o1+ 1-2-1+1+	170 237	0 274 2	32D 30H 27E 26E	496 491 475 474	2	1 1	N20 N35 N27	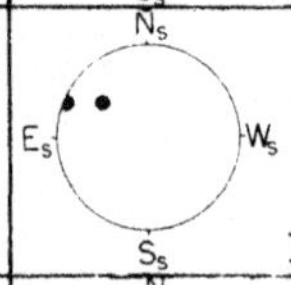	21 15.5
⑲ q			12 0.7 3-4-3+3- 3o2+2o1o	175 230	2 277 2	27^{G}_{H} 29^{G}_{H} 32D 32D	497 494 482 482		1	N10 N25	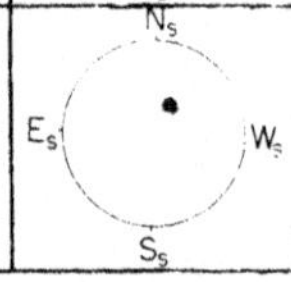	21 13

Feb.10 One of five quiet magnetic days.

Feb.11 Moderate-to-severe magnetic storm with sudden commencements at 0318 and 0756 UT. Outstanding aurora. Dark surge disk, 0605-0758 UT, 06N 60E.

Feb.12 Moderate magnetic activity. Outstanding aurora.

Feb.13 Slight-to-moderate magnetic activity. Outstanding aurora. Spectral radio emission IV, 0000-0009 UT.

Feb.14 Moderate-to-severe magnetic disturbance with sudden commencement 1142 UT. Outstanding aurora. One of five disturbed World Blackout days.

Feb.15 Moderate-to-severe magnetic disturbance continues with resurgence at 2350 UT. Outstanding aurora. One of five disturbed World Blackout days.

Feb.16 Moderate-to-severe magnetic disturbance continues with another resurgence at 0649 UT - one of five disturbed days. Very outstanding aurora. One of five disturbed World Blackout days.

Feb.17 Moderate magnetic activity subsides. Outstanding aurora. U.S.A. launched 1959 alpha (Vanguard II); temperatures and cloud cover; expected life 10 years or more.

Feb.18 One of five quiet magnetic days. Spectral radio emission II, 0508-0523 UT; II, 2241.4-2251 UT.

Feb.19 Slight magnetic activity - one of ten quiet days. Active filament region 2050-2120 UT, 276° position angle, 0.3 radius vector.

Hα flares

Feb.10 - 2, 0216-0256, N12 E74, 1; 2, 0810-0900, N09 E73, 6;
2, 1328-1410D, N09 E70, 4; 2, 1435-1520, N05 E70, 2

Feb.11 - 2, 0605-0805D, N05 E59, 5

Feb.12 - 3, 2301E-0115, N13 E48, 2

Feb.14 - 2, 0926E-0949D, N10 E26, 2

Feb.17 - 2, 0851-0905, N13 E65, 5; 2, 2136-2238, N12 W19, 1

Feb.18 - 3, 1005E-1020D, N21 E76, 1; 2, 1037-1125, N16 E32, 11;
3, 1142-1150D, N21 E76, 1

Feb.19 - 3, 2032-2244, N11 W10, 3

1959

FEBRUARY	GEO-ALERT	SWI	MAGNETIC INDICES Ap C STORM BEGINS ← 3 HR Kp →	R_z 3000 S	IONO-SPHERE INDEX	AUR-ORAL INDEX	COS-MIC RAY INDEX	TYPE IV OUT-BURST	SWF IMP ≥ 2 < 2	ACTIVE REGION LAT SUNSPOT LAT AREA	FLARES POSITION IMP ≥2	HRS. PATROL I_{AREA}
20 Q			4 0.1 0o1o2-2o 1+1-1-1-	160 236	1 258 0	29D 29^{G}_{H} 30E 29E	498 497 487 488	Har		S15	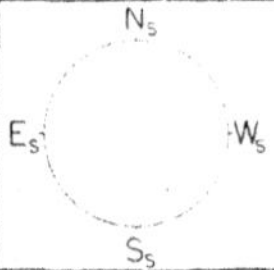	23 10
21 Q			6 0.2 1-0+3-2+ 2o1+2o1+	163 237	1 274 1	26E 31G 34E 34E	500 496 491 493			N 18 N 11 *N 17* *1550* *N 10* *540*	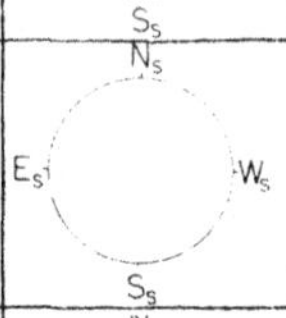	20 12
22			17 0.8 0046 (62) 3o4+3+4o 2+3o2o3o	171 228	1 294 1	29^{G}_{H} 28^{G}_{H} 28^{D}_{E} 31D	501 497 495 496			N28	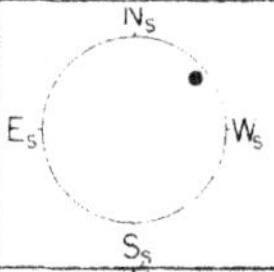	20 10.5
23 q			15 0.3 4-4-4+4- 3o1-1-0o	186 213	3 287 2	33H 30^{E}_{G} 30E <33E	501 499 480 496				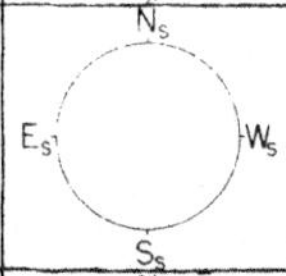	23 10
24 Q			4 0.2 2319 (40) 0o0+1o2- 2-1o1-2-	190 221	0 294 1	 32D 26D 37D	500 498 498 498			N21 N18	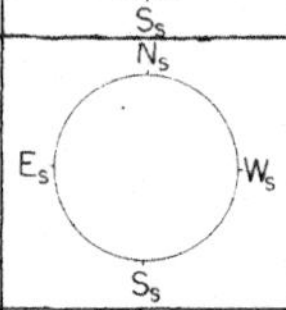	24 12
25 D	MAG STORM AUR 25/ 0215	S	69 1.9 0126 (3) 0215 1157 (13) (3) 3+5-6+6- 6+7+5+4+	181 231	4 287 4	28^{G}_{H} 39D 45E 43C	504 502 501 497				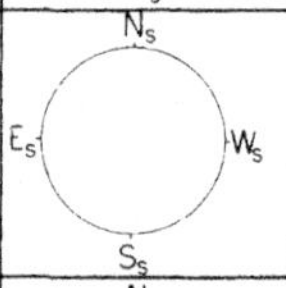	23 16.5
26 D		F	48 1.5 5+5+6+5- 5-5-4+4+	176 217	5 271 5	35D 36D 37E 36E	502 502 500 493			S32 N21 *N23* *1225*	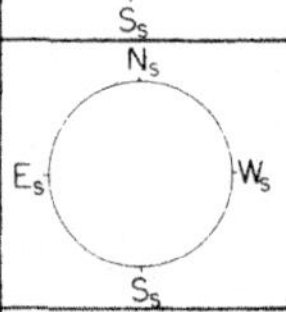	24 13
27			30 1.3 1045 (2) 3+4o3-3o 4+6-5o4-	163 229	3 277 5	33G 31E 39^{D}_{E} 40D	503 504 501 496		1		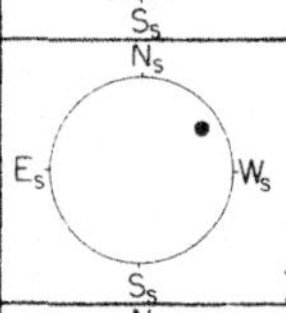	24 16.5
28 D			44 1.5 5-5o3o4o 5o5o6-5+	186 209	5 274 4	34G 34^{E}_{f} 39D 39D	503 505 505 500			N20 N08	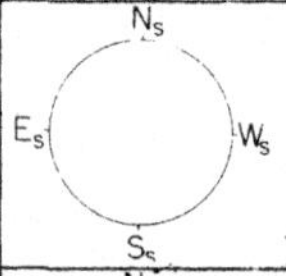	23 12
MAR 1 D			42 1.4 5-5o5o4+ 5-6-4o4+	158 172	5 235 5	36D 34^{E}_{G} 38^{D}_{E} 38^{D}_{E}	504 506 508 500			S06	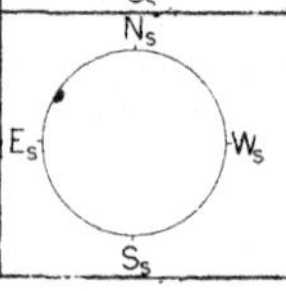	21 8.5

Feb.20 One of five quiet magnetic days. Spectral radio emission II, 1752.5-1804 UT; IV, 1819-1824 UT.

Feb.21 One of five quiet magnetic days.

Feb.22 Slight-to-moderate magnetic activity.

Feb.23 Slight-to-moderate magnetic activity - one of ten quiet days. A quiet day chosen by U.R.S.I. Worldwide Sounding Committee.

Feb.24 One of five quiet magnetic days. A quiet day chosen by U.R.S.I. Worldwide Sounding Committee though sudden commencement at 2319 UT. Outstanding aurora.

Feb.25 Moderate-to-severe magnetic storm with sudden commencements at 0126, 0215 and 1157 UT - one of five disturbed days. A disturbed day chosen by U.R.S.I. Worldwide Sounding Committee. Very outstanding aurora.

Feb.26 Moderate-to-severe magnetic disturbance - one of five disturbed days. A disturbed day chosen by U.R.S.I. Worldwide Sounding Committee. Outstanding aurora. One of five disturbed World Blackout days.

Feb.27 Moderate-to-severe magnetic disturbance continues with resurgence at 1045 UT. Very outstanding aurora.

Feb.28 Moderate-to-severe magnetic activity - one of five disturbed days. Outstanding aurora. One of five disturbed World Blackout days. Sudden disappearance of filament, 0721-1010 UT, 10N 35E.

Mar. 1 Moderate-to-severe magnetic activity continues - one of five disturbed days. Outstanding aurora.

Hα flares

Feb.22 - 2, 0757-0822, N29 W49, 1

Feb.27 - 2, 0740-0840, N23 W46, 10

Mar. 1 - 2, 1812-1845, N32 E90, 1

1959

MARCH	GEO-ALERT	SWI	MAGNETIC INDICES Ap C STORM BEGINS ← 3 HR Kp →	R_z 3000 S	IONO-SPHERE INDEX	AUR-ORAL INDEX	COS-MIC RAY INDEX	TYPE IV OUT-BURST	SWF IMP ≥ 2 < 2	ACTIVE REGION LAT. SUNSPOT LAT. AREA	FLARES POSITION IMP ≥2	HRS. PATROL I AREA
2			31 1.2 4o4+5-4o 4+4-5-4o	151 174	5 230 4	35C 35E 38E 35D	503 504 509 505		1 1	S09 N08 N20	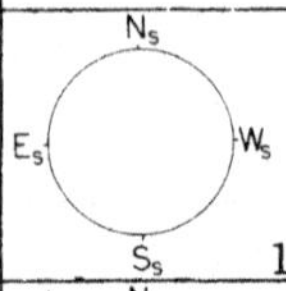	23 18
3			23 1.1 4-4+4-4- 3+4-3+4o	151 139	4 244 4	35D 31H 36D 35 D/E	502 504 509 506		1	N10	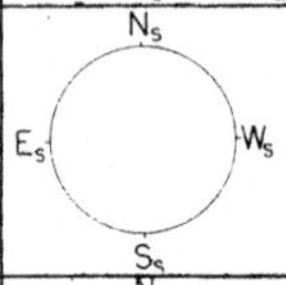	23 9
4			14 0.9 3+2+3o3+ 3-3-3+2o	146 154	3 265 3	35D 33f 33 D/E 35D	503 505 511 509		2	N17 N12	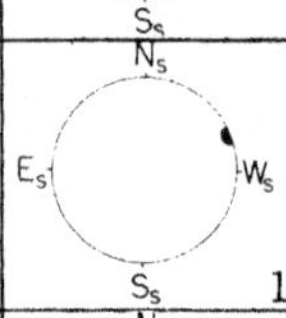	22 10.5
5			12 0.8 3o3o3o2o 2-3-3+3-	152 159	4 250 3	32C 33H 33E 35D	503 504 509 510		2	N15 N23 N12 S17	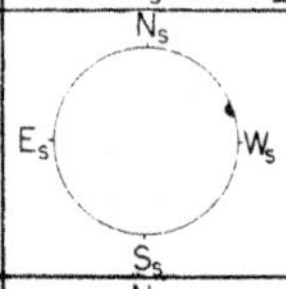	23 9
6 q			6 0.1 3-3-2-1o 1+1-1o0+	148 159	2 278 2	31D 30E 34 C/D 34 C/D	502 503 510 508		1	S17 N17 N25	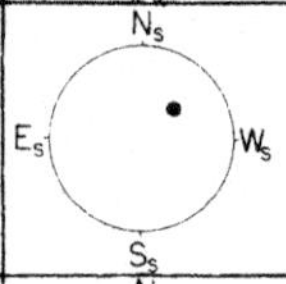	23 6.5
7			6 0.4 0+1-1o2o 2-2o2+3-	150 171	1 237 2	27G 32E 36E 38D	503 503 510 510		1 1	S11 N10 *S11* *500*	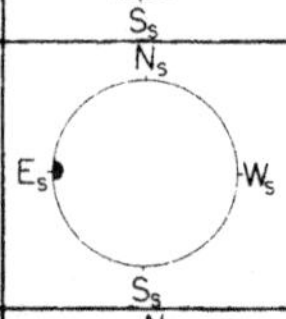	19 3.5
8			11 0.6 5-3+2o1+ 1o2-1-1o	155 183	1 263 3	35D 32E 34D 33E	504 504 510 507				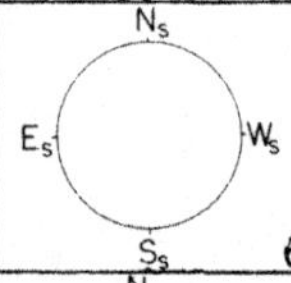	24 6
9 Q			4 0.1 2-2-1+1o 1o2-0+0o	164 183	0 302 2	32H 30G 34D 34D	506 505 510 509		1	N25	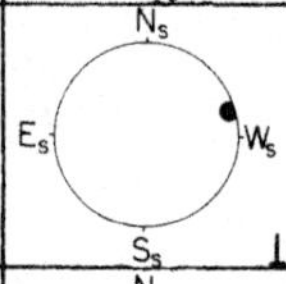	24 10.5
10 Q			2 0.1 0o0o2-2- 1-0+0o0+	156 177	0 299 1	31H 32E 32E 31C	506 507 512 511		2	N23	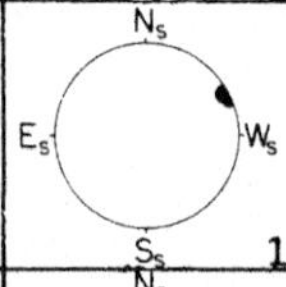	21 11
11 Q			4 0.2 0+0o1+1+ 2-1-2+1+	148 185	1 (312) 1	27 E/G 31D 35D	503 504 509 507	2	2	N10	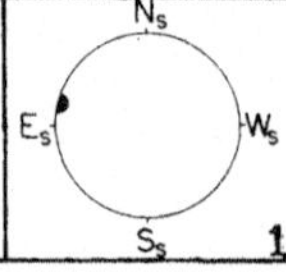	22 11.5

Mar. 2 Moderate magnetic activity continues. Outstanding aurora. One of five disturbed World Blackout days.

Mar. 3 Slight-to-moderate magnetic activity. Outstanding aurora. U.S.A. launched space probe Pioneer IV; corpuscular radiation; passed within 60,000 km of moon, now in solar orbit.

Mar. 4 Slight magnetic activity. Outstanding aurora.

Mar. 5 Slight magnetic activity. Outstanding aurora.

Mar. 6 Very slight magnetic activity - one of ten quiet days.

Mar. 7 Very slight magnetic activity. Outstanding aurora. Active prominence region, 1725-1757 UT, 080° position angle.

Mar. 8 Slight magnetic activity. Outstanding aurora.

Mar. 9 One of five quiet magnetic days - one of quietest days in IGC. Spectral radio emission II, 1723.8 - 1727 UT.

Mar.10 One of five quiet magnetic days - one of quietest days in IGC.

Mar.11 One of five quiet magnetic days. Outstanding aurora. Bright surge limb, 1825-1855 UT, 078° position angle.

Hα flares

Mar. 4 - 2, 0345E-0357, N22 W85, 1

Mar. 5 - 2, 1240E-1257D, N22 W90, 1; 2, 1345E-1402D, N22 W90,1

Mar. 6 - 2, 1225E-1416D, N14 W21, 4

Mar. 7 - 2, 1723E-1750D, N01 E90, 1

Mar. 9 - 2. 1715-1757, N14 W67, 2

Mar.10 - 2, 0704-0755, N26 W70, 2; 2, 1050-1120, N23 W77, 3

Mar.11 - 2, 1803-1847, N14 E82, 3

1959

MARCH	GEO-ALERT	SW	MAGNETIC INDICES Ap	C	3 HR K_p	STORM BEGINS	R_Z 3000 S	IONO-SPHERE INDEX	AUR-ORAL INDEX	COS-MIC RAY INDEX	TYPE IV OUT-BURST	SWF IMP ≥2 / <2	ACTIVE REGION LAT. / SUNSPOT LAT. AREA	FLARES POSITION IMP ≥2	HRS. PATROL / I_{AREA}
12			18	1.0	3+3+3+4- 4o3+3+1+		126 177	1 313 2	35D 38e 34E 34^D_E	503 503 511 507		1	N08	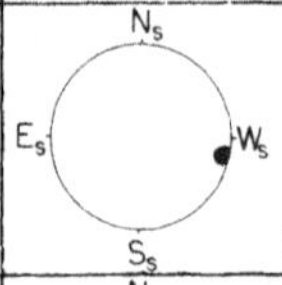	22 2.5
13			12	0.7	2o3o2+3- 3+2+2-3+		159 166	1 293 2	30^C_{DG} 32E 33E 34D	503 503 506 504				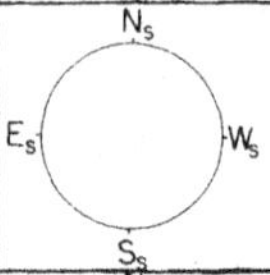	23 7
14			8	0.4	1-2o3o3- 2+1+1o2-		173 197	0 C 1	32C 30E 31E 33D	503 502 506 502		1		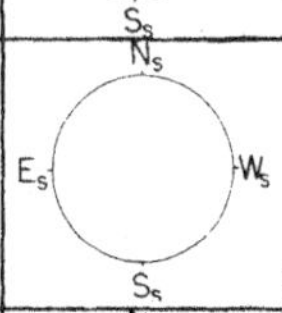	24 8
15			6	0.2	3-2o1+2+ 2-1-0+2-		216 182	1 314 1	30C 28E 28E 30^C_D	502 504 507 507		1	N09 N23 S13 S06	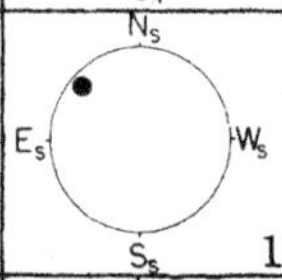	24 11.5
16 Q			4	0.1	2+1o1+1o 1-0+1-1o		225 225	0 (318) 0	30C 28E 29^D_E 29D	504 504 508 508		3 3		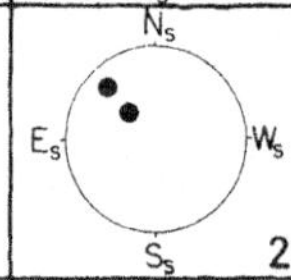	23 23
RWI (17) q			6	0.2	1+2+2o1o 1o2-1o2-		228 204	0 (282) 0	28H 28G 27^E_f 27f	504 506 511 509	1	2	N13 *N11* *1250*	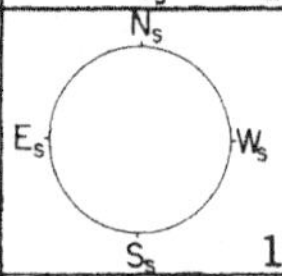	23 15
RWI (18)			7	0.4	0+2o1-1o 2-1+1-4o		230 240	0 319 1	- <28E 28D 27^C_D	505 508 510 508		2	S18 N26 *N25* *2250*	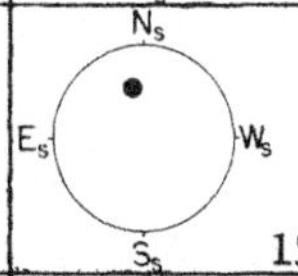	24 15.5
RWI (19) q			6	0.2	1o1o1+2+ 2o2-2+1-		234 253	0 312 1	- - 28E 28E	504 506 510 510		1 1	N31	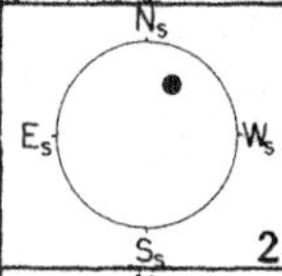	23 24.5
RWI 20 q			5	0.1	1+2o2o1+ 2-1o1o1-		238 260	0 (313) 1	22H 25G 27E 23D	505 505 509 511		1	N19 *N17* *780*	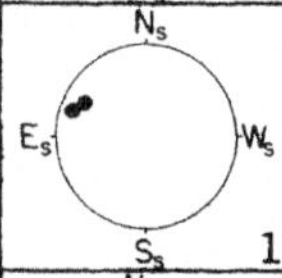	24 15
RWI 21 q			6	0.2	1o1-1+2+ 2o2o1+1+		230 273	0 352 2	28H - 30D 29D	505 506 510 509		3 2	N11 N16 N30	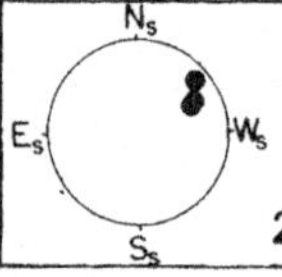	24 22.5

Mar.12 Slight-to-moderate magnetic activity. Outstanding aurora.

Mar.13 Slight magnetic activity. Spectral radio emission II, 0441-0457 UT. U.S.A. rocket, NN3.24F; UV and xrays; 198 km; 1545 UT; White Sands.

Mar.14 Slight magnetic activity.

Mar.15 Slight magnetic activity.

Mar.16 One of five quiet magnetic days —one of quietest days in IGC.

Mar.17 Very slight magnetic activity - one of ten quiet days. Dark surges on disk, 0846-0903, 1235-1240 and 1604-1628 UT, 30N 07E.

Mar.18 Slight magnetic activity. Japanese rocket Kappa - VI-TW-6; atmospheric temperature and wind velocity (grenades); 0505 UT; Akita.

Mar.19 Very slight magnetic activity - one of ten quiet days. Dark surges on disk, 1017-1101 and 1103-1157 UT, 30N 18W.

Mar.20 Very slight magnetic activity - one of ten quiet days. Japanese rocket Kappa VI-TW-7; atmospheric temperature and wind velocity (grenades); 0512 UT; Akita.

Mar.21 Very slight magnetic activity - one of ten quiet days. Active filament region, 0325-0458 UT, 06S 45W.

Hα flares

Mar.12 - 2, 0908-0954, S12 W68, 6

Mar.15 - 2, 1214E-1300, N31 E48, 5

Mar.16 - 2, 0907-1006, N11 E17, 10; 2, 0926E-0937, N29 E37, 1

Mar.18 - 2, 0626-0728, N27 E07, 4

Mar.19 - 2+, 1035-1130, N28 W18, 14

Mar.20 - 2, 0429-0535, N14 E56, 2; 2, 1532-1655, N17 E45, 7

Mar.21 - 2, 0158-0259, N12 W37, 4; 2, 0903-1043, N15 W42, 16;
2, 1645-1712, N29 W49,3

1959

MARCH	GEO-ALERT	SWI	MAGNETIC INDICES Ap	C	STORM BEGINS / ← 3 HR K_p →	R_Z / 3000S	IONO-SPHERE INDEX	AUR-ORAL INDEX	COS-MIC RAY INDEX	TYPE IV OUT-BURST	SWF IMP ≥2 / <2	ACTIVE REGION LAT. / SUNSPOT LAT. AREA	FLARES POSITION IMP ≥2	HRS. PATROL / I_{AREA}
RWI 22 Q			5	0.2	2228 (41) 1-1-1o2- 2o1+1o2o	208 256	0 372 1	- - 25D 27D	505 509 512 509		2		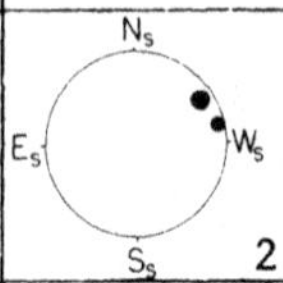	24 21
RWI 23			11	0.7	3-4-3o2- 2-2+3-2-	194 234	0 346 2	28G 28H 28E 27D	504 508 510 508		3	N18	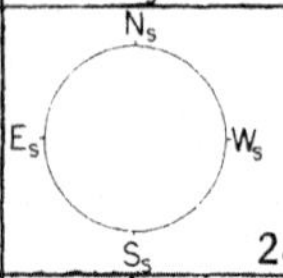	24 28
RWI 24			8	0.5	2o2-2o3- 2+2o2o1+	178 240	1 (361) 1	26G 30E 31D 29D	504 508 510 511	1	2 2	N19 N25 *N 17* *840*	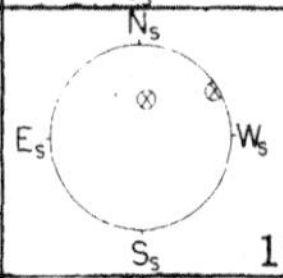	24 19.5
RWI 25			31	1.3	0139 (9) 1204 (15) 2+5+3+2+ 4+5+5o3+	199 212	2 351 3	29H 32G 37f 36D	504 508 507 509		2 1	N18 N24	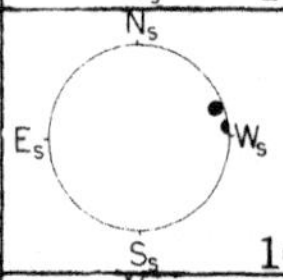	20 16
RWI 26 D	MAG STORM AUR 26/ 0840	S	81	1.8	0842 (71) 2+1+4o6+ 8o5+6+7+	190 203	4 311 4	34G 37D 44E 40E	499 502 496 493		4 1	S33 N24	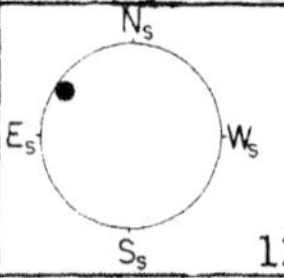	22 12.5
27 D		C	178	1.9	7+8-8+7- 8+8+8-6+	178 208	7 245 5	37C 38E 46D 42D	496 501 488 480			S07 *S08* *570*	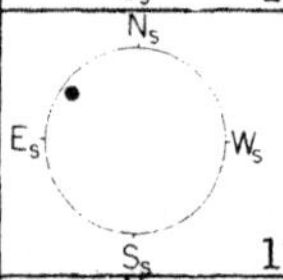	22 12.5
28 D		F	87	1.8	3+5+4+5+ 6+7o7+7+	173 200	6 (222) 5	37C 37E 39$^{D}_{E}$ 42D	498 505 494 485		2	N30 N20	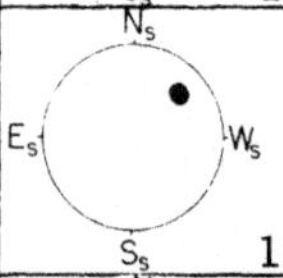	24 15.5
29 D			73	1.6	7-6+6+6- 6-6o5+3-	217 213	6 (238) 5	40D 36H 39D 39D	501 509 504 497		1 1		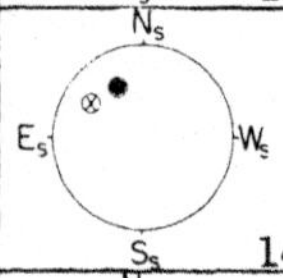	21 14
30			20	1.1	3+4-4+5o 3o2o2o2o	233 190	5 279 4	35C 33$^{E}_{G}$ 37E 35D	503 509 508 502		1	S04 N24	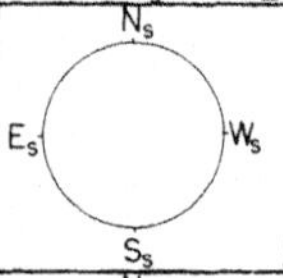	24 5
31			19	1.0	3-3o3-3o 3o2+4o5o	248 239	4 268 2	30C 33E 36D 36D	505 0 511 506			N11 N16 N14 N06 *N25* *1875*	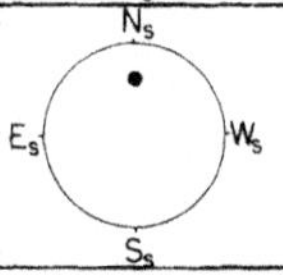	22 9.5

Mar.22 Very slight magnetic activity - one of five quiet days - but sudden commencement at 2228 UT.

Mar.23 Slight magnetic activity.

Mar.24 Slight magnetic activity. Bright surge limb, 0652-0814 UT, 23N 90E.

Mar.25 Moderate-to-severe magnetic disturbance with sudden commencements 0139 and 1204 UT. Outstanding aurora.

Mar.26 Severe magnetic disturbance with resurgence at 0842 UT - one of five disturbed days. Very outstanding aurora.

Mar.27 Severe magnetic disturbance continues - one of five disturbed days one of most disturbed days of I.G.C. Very outstanding aurora. One of five disturbed World Blackout days. Bright surge limb, 2312-2350 UT, 282° position angle.

Mar.28 Moderate-to-severe magnetic disturbance continues - one of five disturbed days. Very outstanding aurora. One of five disturbed World Blackout days.

Mar.29 Moderate-to-severe magnetic disturbance continues - one of five disturbed days. Very outstanding aurora. One of five disturbed World Blackout days.

Mar.30 Moderate magnetic activity. Outstanding aurora. One of five disturbed World Blackout days. U.S.A. rocket AA3.140C; solar physics; 201 km; 1501 UT; Holloman.

Mar.31 Moderate magnetic activity. Outstanding aurora.

Hα flares

Mar.22 - 2, 0403-0421, N24 W52, 1; 2, 2228E-2238, N09 W67, 1

Mar.24 - 2+, 0701-0908, N19 W04, 8; 3, 0958-1325, N29 W77, 16

Mar.25 - 2, 0040-0106, N06 W90, 1; 2, 0615E-0640, N16 W64, 2

Mar.26 - 2, 2100-2147D, N26 E56, 2

Mar.27 - 2, 0128E-0200, N25 E54, 1

Mar.28 - 2, 1725-1840, N23 W34, 2; 3, 2113-2315, N24 W33, 5

Mar.29 - 2, 0747-0930, N17 E37, 6; 2, 1515-1630, N27 E19, 3

Mar.31 - 2, 2130-2225, N32 W01, 3

1959

APRIL	GEO-ALERT	SWI	MAGNETIC INDICES Ap C STORM BEGINS ← 3 HR K_p →	R_z 3000S	IONO-SPHERE INDEX	AUR-ORAL INDEX	COS-MIC RAY INDEX	TYPE IV OUT-BURST	SWF IMP ≥2 <2	ACTIVE REGION LAT. SUNSPOT LAT. AREA	FLARES POSITION IMP ≥2	HRS. PATROL I_{AREA}
1			11 0.6 4-4o4-1o 1o1-1o0+	243 267	2 269 2	31C 32H 30f 34C	505 507 512 505		1		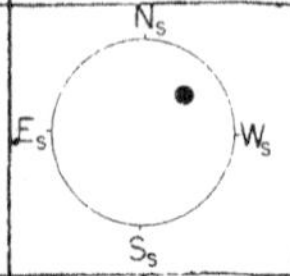	24 9.5
2 q			6 0.3 0+1o1+2+ 1+1+3-2+	242 242	0 280 2	27C 32D 34E 33^{C}_{D}	504 507 511 505				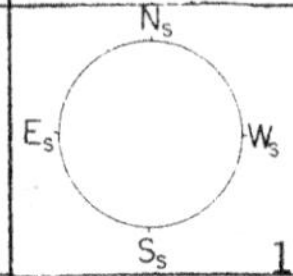	23 13
3			12 0.6 3+3o3-3+ 2+2o1-3-	174 233	1 308 2	28^{G}_{H} 32E 32E 31D	504 507 514 512		1		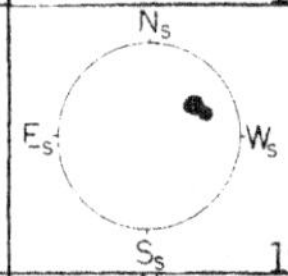	23 12.5
4 q			7 0.2 3-3-2-2- 1-2-1+1o	159 227	1 (275) 0	29H 33E 33E 35C	505 508 517 514		3 2	N15	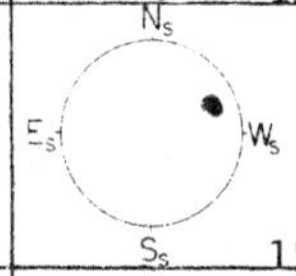	22 15
5 Q			6 0.1 3-2-1o2- 2-2-2-1-	124 208	0 288 0	35C 26G 29E 31D	506 510 519 517		1 2	N16	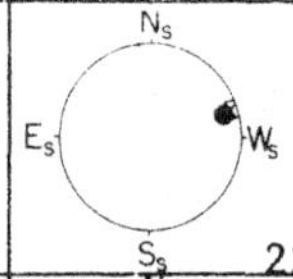	22 22.5
6 q			6 0.2 2o1o1+1+ 1+2o2+2-	108 211	1 310 1	24H 28G 35D 35D	506 509 520 516		2	N13	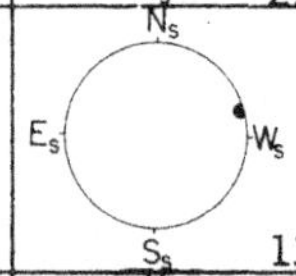	22 13.5
7			8 0.6 2o2-1o2o 2+1+3+2+	107 218	0 (313) 3	29H 30G 35D 35D	504 509 519 521		2 2	N28	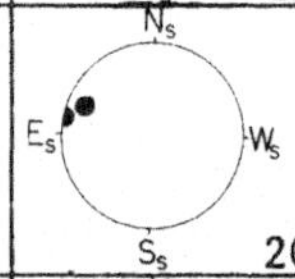	20 20.5
8			22 1.1 3-3+3+3o 4-3+3o5+	136 214	2 304) 3	30H 33G 38^{D}_{E} 38D	507 515 526 527	2	2 1		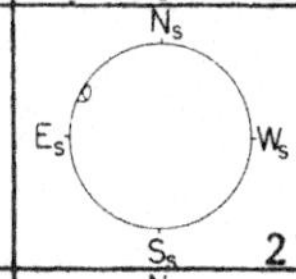	21 21
9 D			44 1.5 1828 (67) 6o6+5-4o 2-2+5+5-	161 220	3 275 4	35C 35G 34D 36C	508 515 529 526	1	2	S22	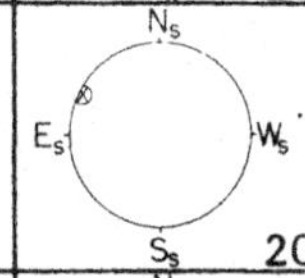	23 20.5
10 D	MAG STORM 09/ 1827	S	98 1.9 0801 (25) 5-5o8-7+ 7+7o5+3+	177 229	5 268 5	37G 35G 41D 42D	503 508 511 507		1	N13	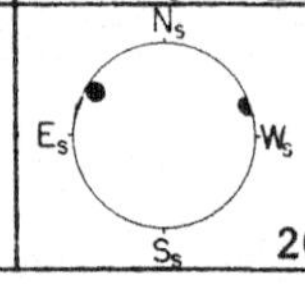	24 20.5

Apr. 1 Slight-to-moderate magnetic activity.

Apr. 2 Slight magnetic activity - one of ten quiet days.

Apr. 3 Slight magnetic activity.

Apr. 4 Slight magnetic activity - one of ten quiet days. Outstanding aurora.

Apr. 5 One of five quiet magnetic days. Sudden disappearance of filament, 0547-0709 UT, 42S 90E.

Apr. 6 Very slight magnetic activity - one of ten quiet days. Outstanding aurora. Bright surges at limb, 0246-0255 and 1321-1350 UT, 20N 90W.

Apr. 7 Slight magnetic activity. Outstanding aurora. Bright surge limb, 1356-1428 UT, 11N 90E.

Apr. 8 Moderate magnetic activity. Outstanding aurora. Bright surge limb, 0915-1000 UT, 28N 90E.

Apr. 9 Moderate-to-severe magnetic disturbance with sudden commencement at 1828 UT - one of five disturbed days. Outstanding aurora. Spectral radio emission II, 1653.4-1703 UT.

Apr.10 Severe magnetic disturbance continues with resurgence at 0801 UT one of five disturbed days. Very outstanding aurora. One of five disturbed World Blackout days.

Hα flares

Apr. 1 - 2, 0455E-0525, N27 W18, 2

Apr. 3 - 2, 0255-0330, N14 W27, 1; 2, 1241-1330, N15 W32, 10; 2, 2210-2240, N10 W38, 1

Apr. 4 - 2+, 0750-0945, N15 W44, 13; 2, 0910-1005, N12 W47, 4

Apr. 5 - 2, 0353-0425, N12 W56, 1; 3+, 2316-0219, N16 W67, 4

Apr. 6 - 2, 1240-1315, N15 W77, 4

Apr. 7 - 2, 0426E-0448, N14 E51, 1; 2, 1355-1535, N10 E84, 3; 2, 1445E-1610D, N15 E48, 1

Apr. 8 - 3, 0905-1010, N27 E85, 10

Apr. 9 - 2, 1645-1710, N25 E70, 2

Apr.10 - 2+, 1036-1110, N26 E54, 5; 2, 1105E-1128D, N18 W72, 1; 2, 1335E-1345D, N17 W75, 1

1959

APRIL	GEO-ALERT	SWI	MAGNETIC INDICES Ap C STORM BEGINS / ← 3 HR K_p →	R_Z / 3000S	IONO-SPHERE INDEX	AUR-ORAL INDEX	COS-MIC RAY INDEX	TYPE IV OUT-BURST	SWF IMP ≥2 / <2	ACTIVE REGION LAT. / SUNSPOT LAT. AREA	FLARES POSITION IMP ≥2	HRS. PATROL / I_{AREA}
11		F	23 1.3 1007 (3) 4-3+4-4o 3o2+4-5o	197 231	2 278 3	36C 34E 40D 42E	502 505 507 511	2	3	N22	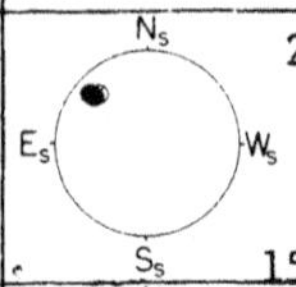	22 15.5
12			10 0.5 4-3o2o3- 2-1+1-3-	189 226	1 261 2	33H 31H 31f 32C	501 507 508 512	1	1 1	N01	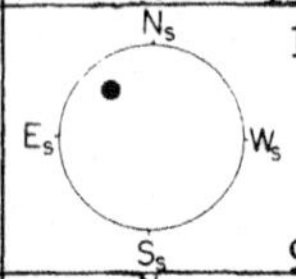	19 9.5
13			9 0.5 3-2o3o2- 3-2o2+1+	178 213	0 304 1	31G 32^{E}_{G} 32E 32D	500 507 510 512	1	1 1	N03 S23 N10	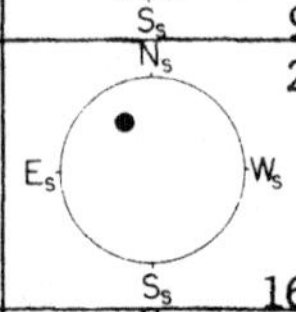	23 16
(14)			11 0.7 4-3o2+2- 2o2o2o3o	193 205	1 294 2	31H 30H 28d 30^{C}_{D}	500 507 510 512	Har	2	S16 *N25* *810*	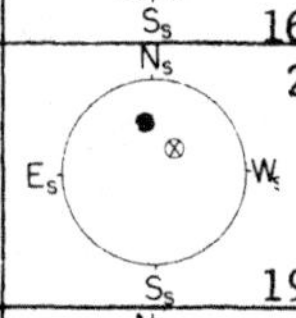	23 19
(15)			10 0.6 4o2-2o1o 1+2o3o2-	174 186	1 271 2	30H 33H 13d 33D	500 508 510 510		1	N27 S09	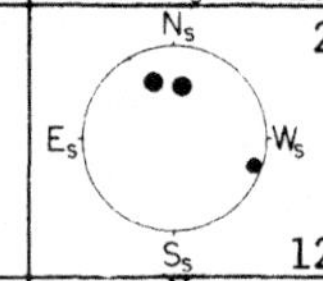	24 12
WMI (16) q			6 0.2 2+1+2-3- 1o1o2-1+	153 183	1 293 2	33H <29E 13d 33D	500 505 510 512			N35 S02	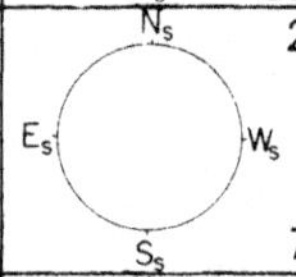	22 7
WMI 17 q			7 0.4 0+1o3-3- 3-1+2-2-	126 185	1 (304) 2	27G 27H 28E 28f	501 503 514 515			N18 N12	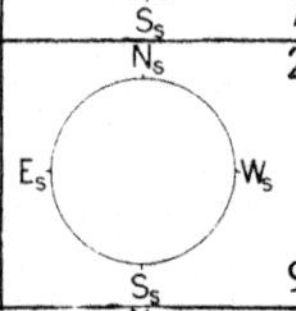	24 9
WMI 18 Q			5 0.2 2+1-1-1+ 1o2-2-0+	114 182	0 269 2	19a - - -	502 503 516 516		1	N19	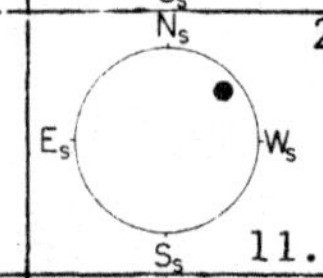	24 11.5
WMI 19 Q			4 0.2 1-1-1-1+ 1+2o1-2o	117 203	0 320 1	- 27H - -	504 506 520 521		1		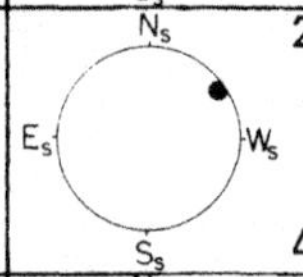	23 4.5
WMI 20 Q			5 0.1 1-1+1o2- 2o1+1o1-	135 208	0 321 1	- - <27D -	502 507 519 502		2 1	N08	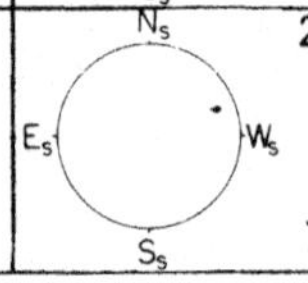	22 7

Apr.11 Moderate magnetic disturbance continues with minor resurgence at 1007 UT. Very outstanding aurora. Sudden disappearance of filament, 0848-1100 UT, 38S 90W. Spectral radio emission II, 2149.4-2154 UT.

Apr.12 Slight-to-moderate magnetic activity.

Apr.13 Slight magnetic activity.

Apr.14 Slight magnetic activity. Spectral radio emission IV, 1824-1831 UT.

Apr.15 Slight magnetic activity.

Apr.16 Slight magnetic activity - one of ten quiet days.

Apr.17 Slight magnetic activity - one of ten quiet days.

Apr.18 One of five quiet magnetic days.

Apr.19 One of five quiet magnetic days.

Apr.20 One of five quiet magnetic days. Active filament region, 0004-0418 UT, 26N 05W.

Hα flares

Apr.11 - 2, 0822-0905, N27 E46, 12; 2, 1446-1607, N26 E41, 8, 2, 2325-0002D, N28 E38, 1

Apr.12 - 2+, 1104-1225, N26 E30, 10

Apr.13 - 3, 0830-0935, N27 E19, 16

Apr.14 - 2, 1220-1245, N27 E05, 7; 2, 1824-2002D, N09 W14, 2

Apr.15 - 2, 0833-0858, N28 W06, 7; 2, 0835-0851, N33 E15, 7; 2, 1050E-1130D, S19 W78, 4

Apr.18 - 2, 0805-0850, N29 W46, 8

Apr.19 - 2, 0823E-0859, N29 W61, 4

1959

APRIL	GEO-ALERT	SWI	MAGNETIC INDICES Ap C STORM BEGINS ← 3 HR K_p →	R_Z 3000 S	IONO-SPHERE INDEX	AUR-ORAL INDEX	COS-MIC RAY INDEX	TYPE IV OUT-BURST	SWF IMP ≥2 <2	ACTIVE REGION LAT. SUNSPOT LAT. AREA	FLARES POSITION IMP ≥2	HRS. PATROL I_{AREA}
WMI 21			8 0.4 3-2+2+3o 2o2+1-0+	128 198	1 301 1	32H 29H - <35C	504 510 519 521			N18 S11		23 7
WMI 22 Q			3 0.0 1-0+0+1- 1-1o1-0+	138 192	0 312 1	- - - <32C	506 508 518 522		1	S13		24 4
WMI 23 D			40 1.5 1036 (69) 0o0+0+4o 5-5o7-6+	186 203	3 326 3	34C - 29f 40D	504 507 517 518		1	N20 S05 S08 N13 N03		22 14
WMI 24 D			27 1.4 4-4o4+4o 5-4+2+4o	186 211	3 283 5	33H 27G 36E 36C	498 499 506 503			N22		24 9.5
WMI 25			22 1.1 3o4-4o3- 3o4o4+4-	203 214	4 (295) 4	35C 30f 37E <33C	500 502 510 508			N23 N18		24 18
26			18 1.1 4-4o3-3o 2o3o3o4+	204 217	3 297 3	33H 32H 30f 33D	502 503 515 510			N09 N23		23 8
27			14 0.9 3o3-3+2o 3o3o2-4-	175 205	3 (278) 3	32H 32H <39D 31C	504 506 519 513	1	1	N15		22 11
28			12 0.9 2-2o3-3o 2+2-2+4+	172 213	2 294 4	33H 30H <40D 33C	504 505 522 517	3	1 2	N25 N11		24 18
29 D			24 1.2 4o4-3o3o 4+5o4o2+	160 209	3 287 3	30H 32H 43D 43D	503 507 521 516		1	N16 N22		23 9
30			21 1.1 2o2o3o2o 5-5o5-2o	141 211	3 295 3	30H 30G 34c 32c	503 506 519 528			N12 N31		21 3

Apr.21 Slight magnetic activity. U.S.A. rocket AA3.181C; ionospheric physics; 235 km; 0105 UT; Holloman.

Apr.22 One of five quiet magnetic days - one of quietest days in IGC. Active prominence region, 0058-0155 UT, 30N 90W.

Apr.23 Moderate-to-severe magnetic storm with sudden commencement at 1036 UT - one of five disturbed days. Very outstanding aurora.

Apr.24 Moderate magnetic disturbance continues - one of five disturbed days. Outstanding aurora.

Apr.25 Slight-to-moderate magnetic activity. Outstanding aurora. One of five disturbed World Blackout days. Sudden disappearance of filament, 1203-1241 UT, 18S 90W.

Apr.26 Slight-to-moderate magnetic activity.

Apr.27 Slight magnetic activity. One of five disturbed World Blackout days.

Apr.28 Slight-to- moderate magnetic activity.

Apr.29 Moderate magnetic activity - one of five disturbed days. Very outstanding aurora. One of five disturbed World Blackout days.

Apr.30 Moderate magnetic activity. One of five disturbed World Blackout days. Sudden disappearance of filament, 1315-1400 UT, 65N 90W.

Hα flares

Apr.24 - 2, 0750-0908, N13 E90, 4

1959

MAY	GEO-ALERT	SWI	MAGNETIC INDICES Ap C STORM BEGINS ← 3 HR K_p →	R_Z 3000S	IONO-SPHERE INDEX	AUR-ORAL INDEX	COS-MIC RAY INDEX	TYPE IV OUT-BURST	SWF IMP ≥2 <2	ACTIVE REGION LAT. SUNSPOT LAT. AREA	FLARES POSITION IMP ≥2	HRS. PATROL I AREA
1			9 0.5 1+4-2o2+ 3-1+2o2o	120 201	2 (288) 4	32G 29H 32c -	503 506 519 514		1	N15 S08	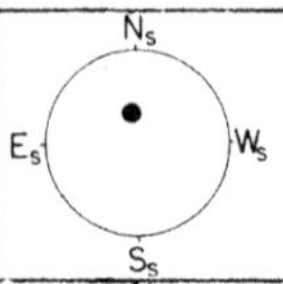	22 8.5
2 q			7 0.3 2o3-2+1o 1o1-2-3-	112 194	1 256 2	32H 30H - -	504 507 520 515		1 3		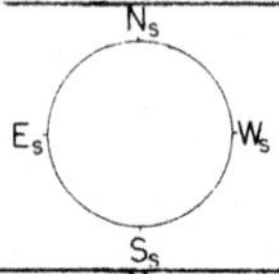	22 8
3 q			7 0.4 1o1o1o1o 1+1+2-4-	113 191	0 (268) 2	19a 26H 32c 32c	505 505 517 514	1	1 4		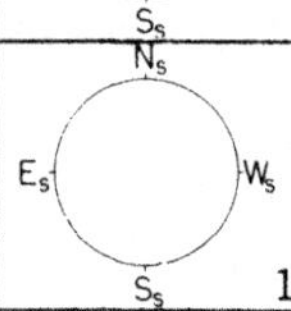	23 13
4			16 1.0 2021 (54) 3o2-2+3o 3-2+3o5o	105 188	1 282 2	34c 31G - 32c	503 501 515 512		4	N06	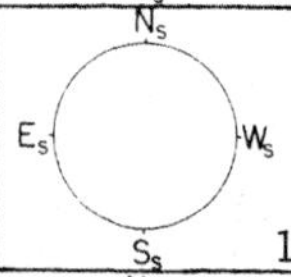	23 19
5			29 1.2 6+5+3o4o 4o2o1o0+	112 186	0 (286) 3	38G 38G 29f -	504 505 516 514		3 6		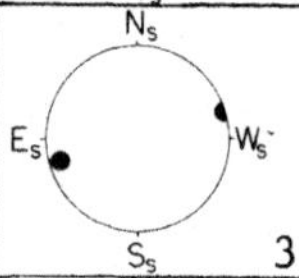	24 39
6 Q			3 0.1 1-1-0+1o 1o1+0+1-	138 191	0 294 0	- 27G - 12d	503 504 514 519		4 6		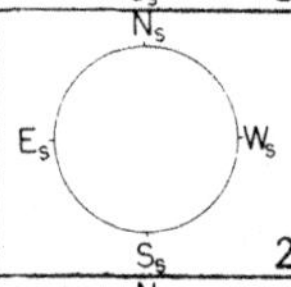	23 20
7 q			6 0.2 2151 (13) 1+1-1o1+ 1o1o1o3o	156 206	0 - 1	- - - 13d	502 503 512 516		1		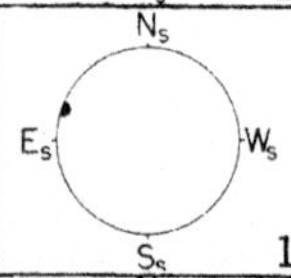	24 19.5
8 D			33 1.3 0353 (3) 3+5+5+5o 4-4o3-4o	188 234	2 (329) 4	33H 35^{G}_{H} 32c 32c	505 507 522 522	1	1 2		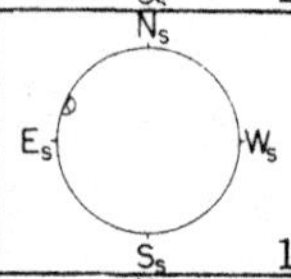	24 16.5
9			16 0.9 3+3-3-4- 4-4-3-3-	238 245	1 319 2	34H 32H 32c 32c	506 509 529 526	1	2	N13 S13 N27 S16	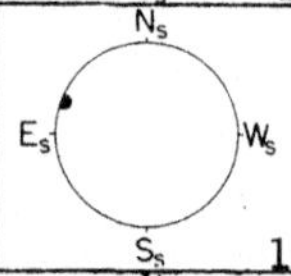	24 19
10			14 1.0 3-2+3+4- 2o3+3o3o	262 246	2 C 2	30^{G}_{H} 32^{G}_{H} 17c 32c	506 508 529 527	Har 1	1 3	N10 *N10* *820*	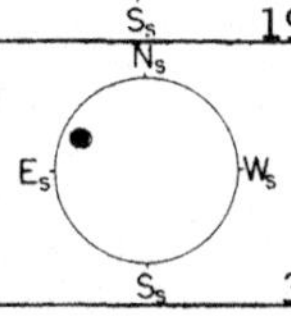	24 32.5

May 1 Slight-to-moderate magnetic activity.

May 2 Slight magnetic activity - one of ten quiet days.

May 3 Slight magnetic activity - one of ten quiet days. Spectral radio emission II, 0011.7-0022 UT.

May 4 Moderate magnetic disturbance begins with sudden commencement 2021 UT. Bright surges at limb, 0853-0923 and 1603-1800 UT, 20N 90E.

May 5 Moderate-to-severe magnetic activity subsides. Outstanding aurora. Bright surges at limb, 0012-0102, 0108-0154, 0142-0213, 0353-0429, 0410-0429, 0822-0841, 1239-1315 and 1343-1424 UT, 25N 90E.

May 6 One of five quiet magnetic days. Bright surges at limb, 0532-0601, 0728-0829, 0928-1114, 1335-1401 and 1500-1545 UT, 25N 90E.

May 7 One of ten quiet magnetic days but sudden commencement of moderate activity at 2151 UT. Bright surges at limb, 1922-1945 and 2046-2120 UT, 25N 90E.

May 8 Moderate magnetic disturbance continues with resurgence at 0353 UT - one of five disturbed days. Outstanding aurora. Bright surges at limb, 0126-0151, 0406-0429, 1031-1048, 1408-1440, 1615-1638 and 1920-1950 UT, 25N 90E. Active prominence region, 2302-2322 UT, 062° position angle. Spectral radio emission II, 2258.7-2319 UT.

May 9 Slight-to-moderate magnetic activity. Bright surges at limb, 0010-0030 and 1520-1620 UT, 28N 90E. Dark surge disk, 1105-1130 UT, 20S 00W.

May 10 Slight-to-moderate magnetic activity. Very large polar cap absorption event, 2300 UT, lasts 7 days (B, RL, K). Dark surges on disk, 2025-2112 and 2032-2110 UT, 23N 53E. Spectral radio emission IV, 2116-2400 UT; II, 2122.8-2141 UT.

Hα flares

May 1 - 2, 0736-0849, N16 E04, 6

May 5 - 2, 0550-0705D, S15 E61,4; 2, 0605-0725D, N16 W82, 5

May 7 - 2, 1307E-1355D, N19 E80, 1

May 8 - 2+, 2252-2346, N21 E83, 3

May 9 - 3, 0123-0312D, N20 E78, 1

May 10 - 2, 2012-2044, N19 E51, 3; 3+, 2102-0210D, N18 E47, 6

1959

MAY	GEO-ALERT	SWI	MAGNETIC INDICES Ap	C	STORM BEGINS	← 3 HR K_p →	R_z 3000S	IONO-SPHERE INDEX	AUR-ORAL INDEX	COS-MIC RAY INDEX	TYPE IV OUT-BURST	SWF IMP ≥2 / <2	ACTIVE REGION LAT. SUNSPOT LAT. AREA	FLARES POSITION IMP ≥2	HRS. PATROL I AREA
11			26	1.3	2328 (48)	3o2+4o3o 4-4-4-6o	287 289	8 (333) 3	32H 32H 30c 32c	506 510 532 530	Har 1	1	S19 N12 *N25* *640* *S18* *780*	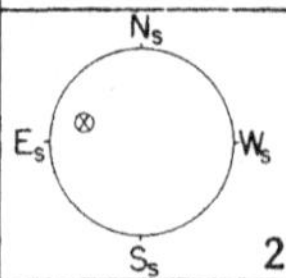	24 21
(12) D	MAG STORM 11/ 2318	S	108	1.9		7o7-7-6- 7-8+5+5-	276 275	9 298 4	34c 36H 46D <46E	483 478 462 465		3 2	N26 N32	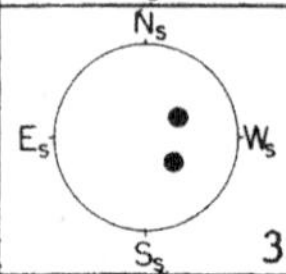	24 30
(13)		F	14	0.8		2-2+4-3- 3-3o3o3+	·257 265	9 (321) 3	- 28f 30f 32c	486 482 469 474	Syd 1	2 2	S26	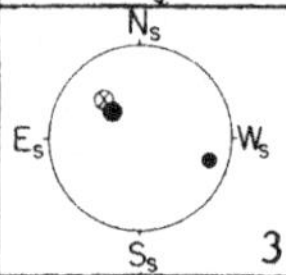	24 31
(14) Q			5	0.2		1+1+0+1+ 3-2-1-1-	227 243	8 (305) 3	18a - 28f -	483 476 461 469			N14 *N14* *640*	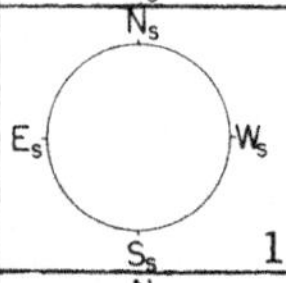	22 19
15 D			33	1.2	0703 (22)	2+2o3-5- 5o5+5-5+	204 227	7 (295) 3	- 29f 31f 34c	487 481 469 472			N18 N08 *N19* *1500*	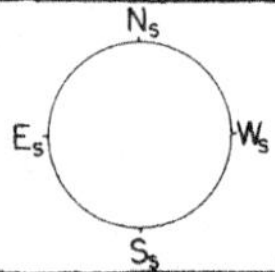	22 3
16 D	MAG STORM 15/ 1100		36	1.3		6o6-6-4- 2+2-2o4+	182 221	8 C 3	35a 36H 28f 28f	490 485 473 474	1	5	N21	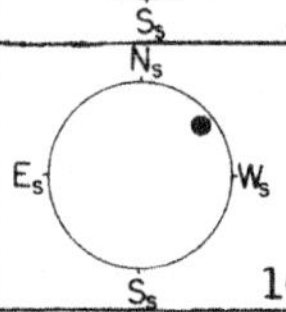	24 16.5
17			13	0.8		2-1+3-3o 4-4o2+2+	187 216	4 C 2	- 29H - -	492 488 479 478	3	3 2	N18	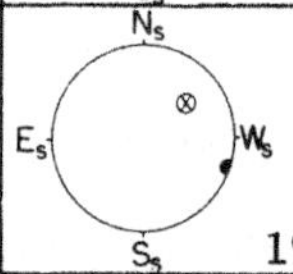	23 19
18			24	1.2		3+5-4o2+ 2+3+4-5o	198 206	3 (349) 2	33G 36G 16c 32c	493 492 484 485	1	1 1	N10	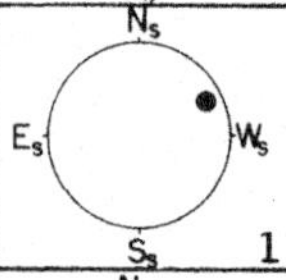	21 11.5
19			14	0.8		3+3o2-3o 3o3-3o3o	185 201	3 - 2	29H - 16c -	497 495 490 494		1 1		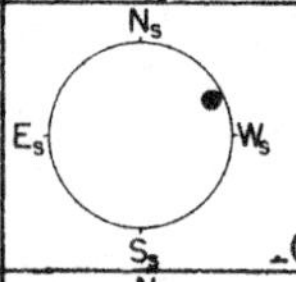	23 0.5
20			10	0.7		3o2-2+2+ 2-2+3+2+	187 201	2 (303) 2	28G 31f - 29f	497 497 493 498		1	S05 N23 N18	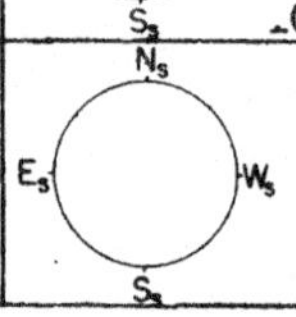	24 12

May 11 Severe magnetic storm begins with sudden commencement at 2328 UT. Forbush cosmic ray decrease, onset before S.S.C. (Uppsala). One of five disturbed World Blackout days. Spectral radio emission IV, 0000-0130 UT; II, 2020.4-2039 UT.

May 12 Severe magnetic storm continues - one of five disturbed days - one of most disturbed days of IGC. Very outstanding aurora. One of five disturbed World Blackout days. U.S.A. rocket AA3.200C; atmospheric composition and upper atmos. physics; 145 km.; 1139 UT; Holloman.

May 13 Slight-to-moderate magnetic activity. One of five disturbed World Blackout days. Dark surges on disk, 0506-0535, 0542-0719, 0750-0815 and 0820-0845 UT, 20N 20E. Bright surge at limb, 0858-0952 and 2340-2418 UT, 08S 90E. Spectral radio emission II, 0525-0630 UT; II, 1424.6-1430 UT.

May 14 One of five quiet magnetic days. One of five disturbed World Blackout days.

May 15 Moderate magnetic disturbance with sudden commencement at 0703 UT - one of five disturbed days.

May 16 Moderate-to-severe magnetic disturbance - one of five disturbed days. Outstanding aurora. One of five disturbed World Blackout days. Dark surge disk, 0923-1048 UT, 10S 50E.

May 17 Slight-to-moderate magnetic activity. Dark surge disk, 0530-0603 UT, 26N 36W.

May 18 Moderate magnetic activity. Outstanding aurora. Spectral radio emission II, 0407-0412 UT.

May 19 Slight magnetic activity. Spectral radio emission II, 1343.8-1353 UT.

May 20 Slight magnetic activity.

Hα flares

May 11 - 3, 2006-2150D, N10 E41, 4

May 12 - 2, 0655E-0801, N10 W22, 6; 2, 1009-1100, S17 W18, 14

May 13 - 2+, 0509-0553, N22 E26, 7; 2, 0855-1042, S10 E90, 5; 2, 1553-1655, N14 E18, 6; 2, 1815E-1915D, S15 W53, 1

May 16 - 2, 0708-0815, N29 W50, 3

May 17 - 2+, 0523-0600, N20 W30, 2; 2, 1558-1629D, S18 W85,1

May 18 - 2, 1546-1625, N20 W53, 8

May 19 - 2, 1332-1401, N21 W57, 4

1959

MAY	GEO-ALERT	SWI	MAGNETIC INDICES Ap C STORM BEGINS ← 3 HR K_p →	R_z 3000 S	IONO-SPHERE INDEX	AUR-ORAL INDEX	COS-MIC RAY INDEX	TYPE IV OUT-BURST	SWF IMP ≥ 2 < 2	ACTIVE REGION LAT. SUNSPOT LAT. AREA	FLARES POSITION IMP ≥2	HRS. PATROL I_{AREA}
21			13 0.8 3o3-2+3- 2-2+4-3+	151 197	1 303 3	29H 28G 31f 29f	496 495 496 501				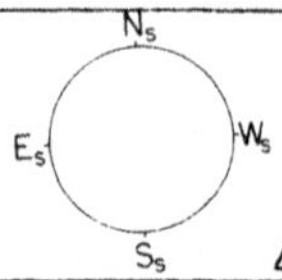	20 4
22			15 0.9 3+2+3+3o 2o4-3o3o	145 199	2 268 3	- 30f 28f -	498 496 500 507				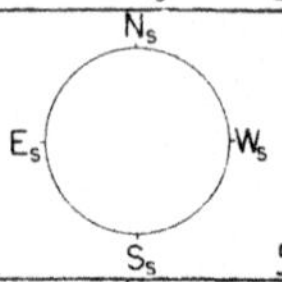	24 5
23			12 0.7 4-3+2+2+ 2-2+3-2+	149 194	1 270 3	30H 28 29f -	498 499 503 510			N23 N10	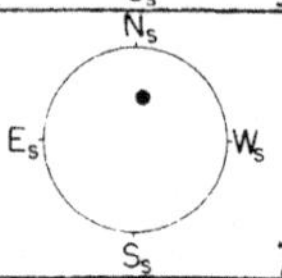	24 7.5
24 D	MAG STORM 24/ 0540		52 1.6 0540 (67) 2027 (20) 3+4o6-5o 5-4o5-7+	143 197	3 248 5	33H 34G 32C -	499 500 504 504		1		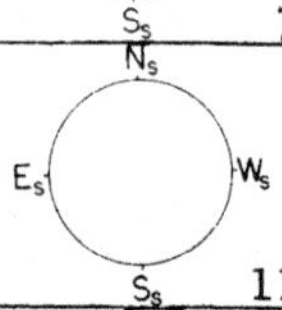	23 11
25			25 1.3 6+4-4-3+ 2+3o3-2+	178 196	1 - 5	32H 33H 34C -	498 499 499 494			N12 N20	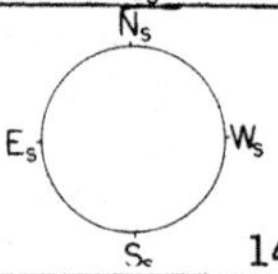	24 14
26 q			7 0.4 3+1o1-2o 1+1+2+2-	188 202	1 291 3	- - - -	500 501 503 498		1	N03 N14 S16 S16 1200	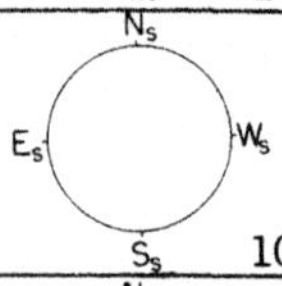	24 10
27 Q			5 0.2 2+2-1+1+ 1o0+1-1o	177 204	1 (274) 2	29H 28G - -	500 504 506 503		3		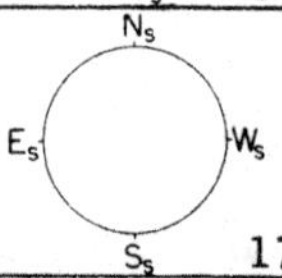	24 17
28 Q			3 0.1 0+1o1-1- 1o0+1-1+	132 190	1 (287) 1	27H 28G - -	503 0 508 507		3	S08 N22	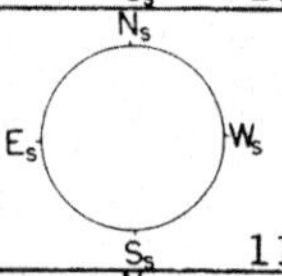	24 11
29 Q			3 0.0 1-0+0+1- 0+1-0+2-	99 174	1 (267) 1	- - - -	503 0 511 514				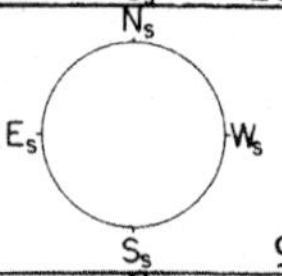	22 9
30 q			6 0.3 1o0+0+2- 3+2o1+1-	106 176	1 (275) 3	19^G_H - 34c 12d	505 - 513 517		1	S09	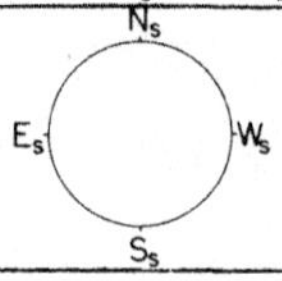	23 5

May 21 Slight-to-moderate magnetic activity. Active filament region, 0025-0130 UT, 30N 02W.

May 22 Slight-to-moderate magnetic activity. U.S.A. rocket AA3.201C; atmospheric composition and upper atmos. physics; 142 km.; 1115 UT; Holloman.

May 23 Slight-to-moderate magnetic activity. Spectral radio emission II, 0101.5-0114 UT.

May 24 Moderate-to-severe magnetic storm begins with sudden commencement at 0540 UT and resurgence 2027 UT - one of five disturbed days.

May 25 Moderate magnetic disturbance subsides.

May 26 Slight magnetic activity - one of ten quiet days. Bright surge limb, 0428-0633 UT, 35N 90W.

May 27 One of five quiet magnetic days. U.S.A. rocket AA6.161C; upper atmos. physics; 169 km.; 0451 UT; White Sands. U.S.A. rocket AA6.162C; upper atmos. physics; 177 km.; 1950 UT; White Sands.

May 28 One of five quiet magnetic days - one of quietest days in IGC.

May 29 One of five quiet magnetic days - one of quietest days in IGC.

May 30 Very slight magnetic activity - one of ten quiet days.

Hα flares

May 23 - 2, 0139-0229, N26 W05, 2

1959

MAY	GEO-ALERT	SWI	MAGNETIC INDICES Ap C STORM BEGINS ← 3 HR K_p →	R_z 3000 S	IONO-SPHERE INDEX	AUR-ORAL INDEX	COS-MIC RAY INDEX	TYPE IV OUT-BURST	SWF IMP ≥ 2 < 2	ACTIVE REGION LAT. SUNSPOT LAT. AREA	FLARES POSITION IMP ≥2	HRS. PATROL I_{AREA}
31			22 1.1 2+3+2+2+ 5o4-4o4+	131 182	1 272 3	31H 29H 34c 34c	504 0 517 520		2	S34	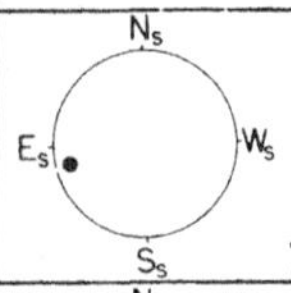	24 7.5
JUN 1 q			7 0.3 3-2o2o1+ 1o2-2+2o	152 184	1 (294) 3	25G 27H 30f 32c	504 509 520 523		2	N33	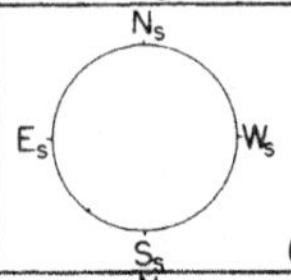	24 6
2			19 1.0 2+3-3o3+ 4o3o4o4+	133 188	0 268 3	33H 32^{G}_{H} 32c 34c	505 507 521 524		1 2		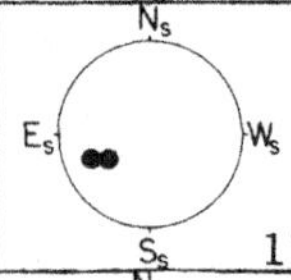	23 11
3			16 1.0 3+4o3-3+ 4-3o2o2+	152 187	1 (262) 4	32H 35G 32c -	505 510 520 522		1		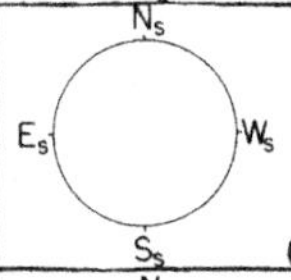	23 6
4 D			22 1.1 4-3o4o4- 3-3-4-5o	166 184	2 - 3	29H 28H - 34c	504 507 516 520			S08 S14 *S12* *950*	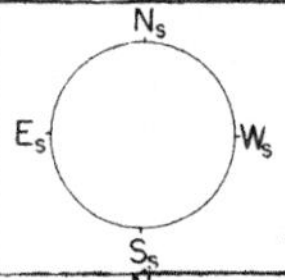	24 7
5			15 0.8 4+3o2+3- 3+3-3-2+	162 185	1 - 3	32H 30H 31f 29f	505 506 516 518			S14	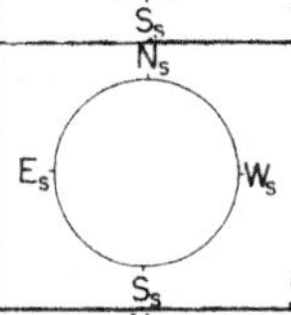	22 5
6			12 0.7 2-2-2+3+ 3-3o3o3o	180 197	1 - 2	20a 29^{e}_{f} 30f 16c	505 507 517 519		1	N07	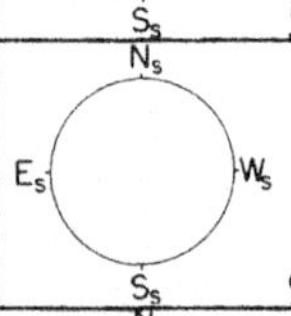	24 9
7			8 0.5 2+3o2-1+ 2o3-2o2-	181 194	0 (267) 2	27H 27G - -	504 506 518 521		1 3	N11	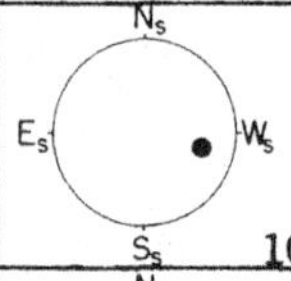	23 10.5
8			11 0.7 2-1+3-2o 3-3+3o3-	192 204	1 264 2	- - 32c 32c	504 507 521 525			N10	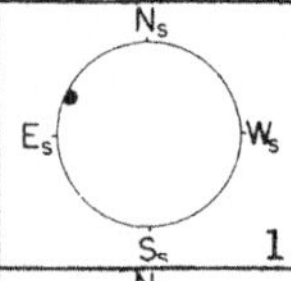	24 11.5
9			19 1.0 0016 (40) 3o3+4o4+ 4o3-3o2+	188 216	2 (303) 4	29^{G}_{H} 33G 29c 29f	504 506 521 523	3	1	S24 *S23* *940*	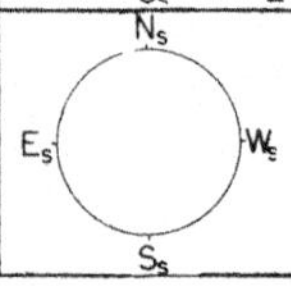	23 7.5

May 31 Moderate magnetic activity.

June 1 Slight magnetic activity - one of ten quiet days.

June 2 Slight-to-moderate magnetic activity.

June 3 Slight-to-moderate magnetic activity. Outstanding aurora. Sudden disappearance of filament, 0730-1230 UT, 19S 43E.

June 4 Moderate magnetic activity - one of five disturbed days.

June 5 Slight-to-moderate magnetic activity. Spectral radio emission II, 0546-0550 UT.

June 6 Slight magnetic activity.

June 7 Slight magnetic activity.

June 8 Slight magnetic activity.

June 9 Slight-to-moderate magnetic activity with sudden commencement 0016 UT. Active prominence region, before 1720-0300 UT, 20N 90E.

Hα flares

May 31 - 2, 1017E-1018D, S13 E59, 1

June 2 - 2+, 1808-1908, S15 E27,5; 2, 2000-2030, S15 E43, 3

June 7 - 2, 0805-0854, S08 W40, 4

June 8 - 2, 0250-0326, N24 E71, 1

June 9 - 2, 1800E-1910, N17 E90, 1

1959

JUNE	GEO-ALERT	SWI	MAGNETIC INDICES Ap	C	STORM BEGINS	← 3 HR K_p →	R_z / 3000S	IONO-SPHERE INDEX	AUR-ORAL INDEX	COS-MIC RAY INDEX	TYPE IV OUT-BURST	SWF IMP ≥2 / <2	ACTIVE REGION LAT. / SUNSPOT LAT AREA	FLARES POSITION IMP ≥2	HRS. PATROL / I_{AREA}
10			12	0.7		4-4+2+2+ 2o2o1+2o	160 224	1 312 3	30H 29G - 18c	503 505 518 515	1	1 2		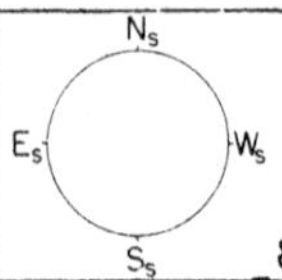	24 8
11	MAG STORM 11/ 0910	S	24	1.1	0909 (55)	1+1+2-5o 6o5-2o2-	172 234	2 357 2	27c 32f 17c 32c	501 502 508 507	1	2 1	N07	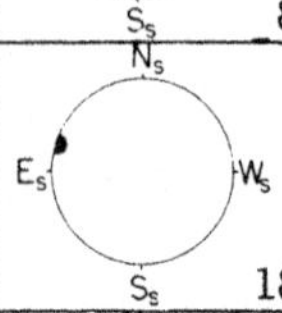	24 18.5
12 Q		F	6	0.2		1o1o1o1- 1o3-3o1+	176 203	1 (324) 3	36H - - 28c	499 501 507 511	2	1	N23 N13	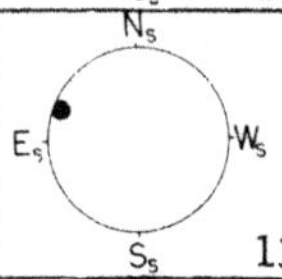	24 13.5
13 Q			4	0.1		1+1-1-1- 1+1-1+2-	165 214	1 - 2	- - - -	503 506 515 521	1	1	N22	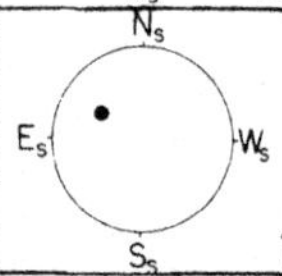	23 7.5
14			9	0.4		1-2-1o2+ 3+2+1+3o	170 209	2 (308) 2	- - 30f -	501 504 517 521		1		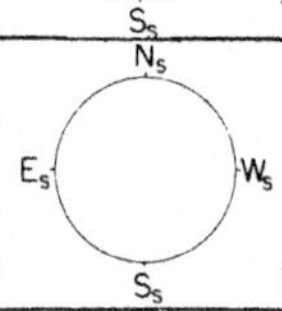	24 7
RWI 15 q			8	0.4		2-2+3+2+ 2o1o1o1+	158 229	3 - 2	- 31G 15d 27d	499 499 510 514		2 3		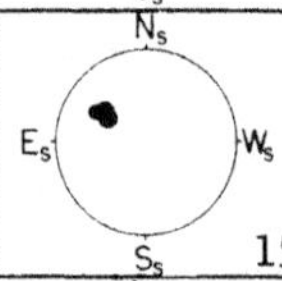	24 15.5
RWI (16) Q			6	0.2		2o2-2-1+ 1+1o2+2+	172 216	2 (317) 2	27H 27H - -	499 500 507 511		2		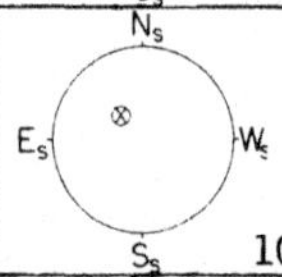	24 10
RWI (17) Q			6	0.3	1444 (3)	0+0+0+0+ 1o1+3-3+	161 221	1 - 0	- - - -	500 500 507 511		1	S20 N18 N28 N17 760	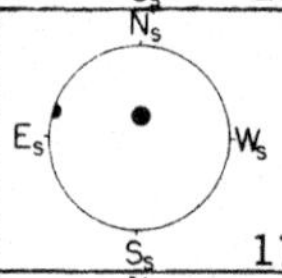	24 17
RWI (18)			8	0.6		2-2+2o1+ 3+2+2o2-	174 222	1 302 1	24H 29G - 13d	499 500 505 510		1 1	S05	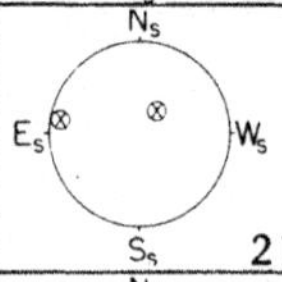	24 21.5
RWI 19 q			7	1.4		1+3o2o2o 2+1o2o1+	182 229	0 - 1	24H 31f - -	0 499 504 510		1	N10 N24	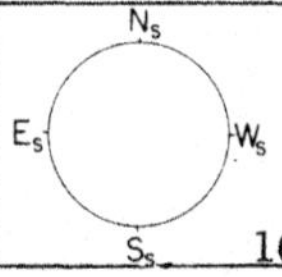	24 16

June 10 Slight-to-moderate magnetic activity. Sudden disappearance of filament, 0625-0744 UT, 20N 90W. Spectral radio emission II, 0247-0308 UT. Solar electron corona, Climax. U.S.A. rocket AA3.280C; geomagnetic field; 235 km.; 1640 UT; Holloman.

June 11 Brief moderate-to-severe magnetic disturbance with sudden commencement at 0909 UT. Bright surges at limb, 0310-0339, 0801-0846, 0945-1008, 1154-1217, 1215-1241 and 1400-1440 UT, 11S 90W; 1804-1845 UT, 15N 90E. Solar electron corona, Climax.

June 12 One of five quiet magnetic days. Outstanding aurora.

June 13 One of five quiet magnetic days - one of quietest days in IGC.

June 14 Slight magnetic activity.

June 15 Slight magnetic activity - one of ten quiet days. One of five disturbed World Blackout days. Bright surge limb, 0438-0628 UT, 16N 90E.

June 16 One of five quiet magnetic days.

June 17 One of five quiet magnetic days. Solar electron corona, Climax.

June 18 Slight magnetic activity.

June 19 Slight magnetic activity - one of ten quiet days.

Hα flares

June 11 - 2, 1802-2130, N17 E78, 5

June 12 - 2+, 0735-1020, N21 E65, 14

June 13 - 2, 1051-1140, N17 E27, 1

June 15 - 2+, 0230-0420, N20 E29, 3; 2, 0520-0825, N19 E26, 10; 2, 1050-1210, N16 E26, 9

June 16 - 3, 0618-0810, N16 E15, 11

June 17 - 2, 0245-0305, N16 E80, 1; 2, 1421-1604, N16 W02, 16

June 18 - 2, 1102E-1312, N08 E61, 12; 3+, 1130-1310, N16 W12, 9

1959

JUNE	GEO-ALERT	SWI	MAGNETIC INDICES: Ap, C, STORM BEGINS / ← 3 HR K_p →	R_z / 3000 S	IONO-SPHERE INDEX	AUR-ORAL INDEX	COS-MIC RAY INDEX	TYPE IV OUT-BURST	SWF IMP ≥2 / <2	ACTIVE REGION LAT. / SUNSPOT LAT. AREA	FLARES POSITION IMP ≥2	HRS. PATROL / I_{AREA}
RWI 20 q			8 0.4 3-2-1+1+ 2+1+2+2+	173 220 2	0 (284) 0	- - 20f	500 499 503 511		1	S08 N15	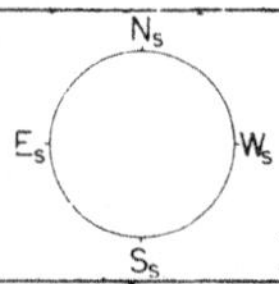	24 10
RWI 21 q			8 0.4 2-1o2-1o 2o2+3o2+	162 218	0 (284) 1	33H 27H - -	499 498 503 510		2		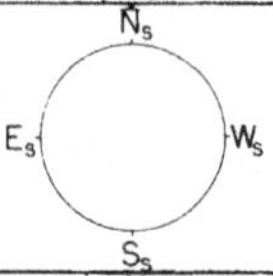	23 9
RWI 22			11 0.8 1651 (3) 2-2-2o3+ 3o3-3-3o	170 213	1 - 2	- - - -	498 498 503 510		1 1	N21 N04 S16 N12 N10 2400	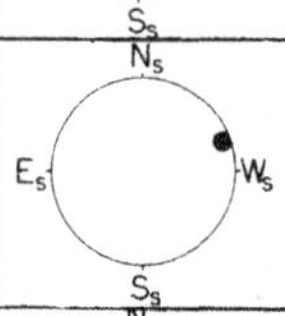	23 7.5
RWI 23			15 0.9 1316 (9) 3-2o2o2- 2-4-4-5-	188 215	0 (296) 2	29A 29G - -	500 500 504 507		1 2	N08 N25	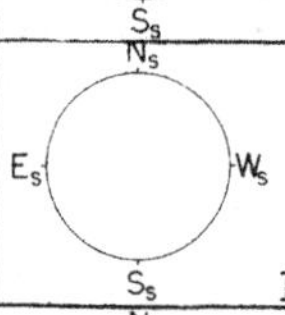	23 17
RWI 24			26 1.1 6-6o3-1+ 1o2-4-3-	157 221	2 (284) 3	36H 32G - 18a	501 501 508 510				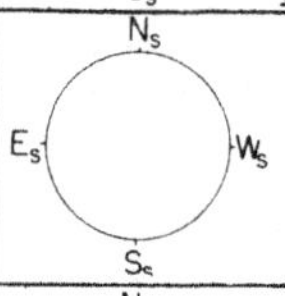	24 6
25 Q			6 0.4 2o1+1o2- 1+1o2-3o	180 223	0 (279) 1	- 26f - 20a	502 499 505 509			N09	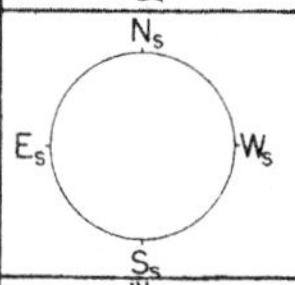	23 8
26			14 0.8 3-3+4+3o 3o3-2-1-	184 222	2 290 3	31H 31H 32c -	503 501 514 512			S08 N15	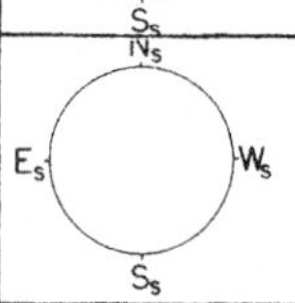	23 7
27 D			31 1.3 0651 (3) 1119 (4) 0+1-2+3o 6o6-4o5o	186 227	3 299 3	30H - 34$^{d}_{f}$ 34c	500 500 511 513		3	N27 N16	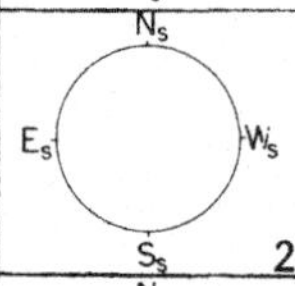	24 22
28 D	MAG STORM 27/ 1100		34 1.3 5o4+4+3+ 6-4o4o3+	160 225	5 279 4	34c 35G 34d 16d	498 499 505 511		1	N33 N20 S13 S17	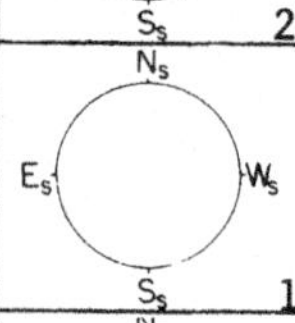	23 11
29 D	MAG STORM 29/ 0728		51 1.5 0728 (49) 3o3+6o6o 5o4+5-6+	158 205	6 - 5	30A 36H 34c 34c	499 501 506 510				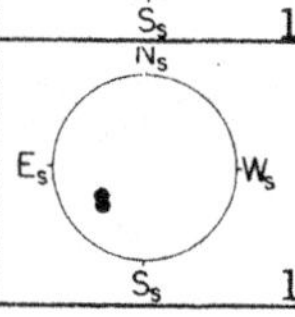	23 12

June 20 Slight magnetic activity - one of ten quiet days.

June 21 Slight magnetic activity - one of ten quiet days.

June 22 Slight magnetic activity with sudden commencement at 1651 UT. Spectral radio emission II, 0503-0512 UT.

June 23 Moderate magnetic disturbance with sudden commencement 1316 UT.

June 24 Moderate-to-severe magnetic disturbance subsides. U.S.A. rocket AA3.220C; ionospheric physics; 200 km.; 1210 UT; Holloman.

June 25 One of five quiet magnetic days.

June 26 Slight-to-moderate magnetic activity.

June 27 Moderate-to-severe magnetic storm with sudden commencements at 0651 and 1119 UT. - one of five disturbed days. One of five disturbed World Blackout days.

June 28 Moderate-to-severe magnetic disturbance continues - one of five disturbed days. Outstanding aurora. One of five disturbed World Blackout days.

June 29 Moderate-to-severe magnetic activity continues with resurgence at 0728 UT - one of five disturbed days. Outstanding aurora. One of five disturbed World Blackout days.

Hα flares

June 22 - 2, 1010-1135, N18 W67, 12

June 29 - 2, 0854E-0954, S20 E29, 8; 2, 1305-1400, S15 E29, 7

1959

JUNE	GEO-ALERT	SWI	MAGNETIC INDICES Ap	C	STORM BEGINS / 3 HR Kp	R_z 3000S	IONO-SPHERE INDEX	AUR-ORAL INDEX	COS-MIC RAY INDEX	TYPE IV OUT-BURST	SWF IMP ≥ 2 / < 2	ACTIVE REGION LAT. / SUNSPOT LAT. AREA	FLARES POSITION IMP ≥2	HRS. PATROL / I_{AREA}
30 D			38	1.4	0801 (29) 5-3o4+5+ 6o5-4o3o	147 194	4 - 5	32$^{c}_{H}$ 33$^{G}_{H}$ 37e 34c	500 502 509 510				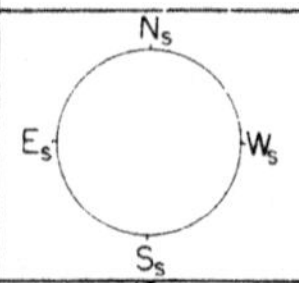	24 3
JUL 1 Q			5	0.2	1+1+2-1- 1o1-1o2+	147 184	1 (274) 2	32H 29H 27f 28c	501 504 509 514		1	N10 S12 S23 N29	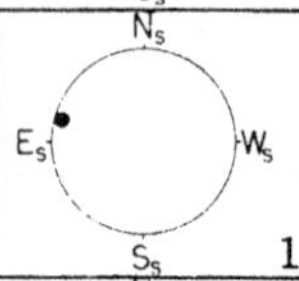	24 12.5
2 q			14	0.8	3+4+4+2+ 2o2o1-1-	118 180	2 (261) 2	33G 34$^{G}_{H}$ 32c 19a	501 508 513 515				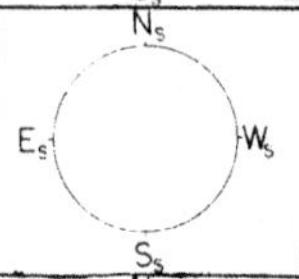	24 6
3 Q			5	0.1	2-1+2-1o 1+1o1+1o	138 173	1 (277) 1	19a - - -	502 506 512 0				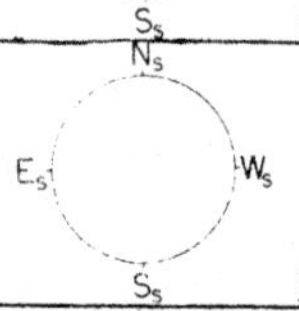	24 5
4 q			15	0.8	1-0+1o3o 3-4o4o4o	158 181	2 - 2	- 28$^{f}_{G}$ 29f 32c	500 498 505 509			N10 N14 *N08* *540* *N14* *1175*	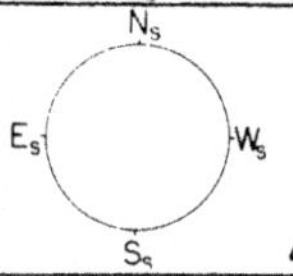	24 4
5			16	0.9	3+4-3+3o 4-2o3o2+	136 180	3 - 3	33H 35G 28f 28c	500 501 512 517			N21 S07	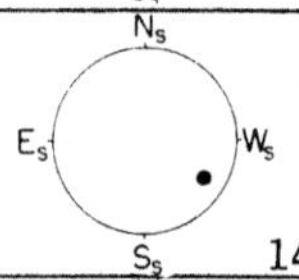	24 14.5
6 q			11	0.6	2-1+2o3- 3+3-2o3-	127 189	2 (256) 2	26H 30G 29f 27h	501 501 512 515			N18	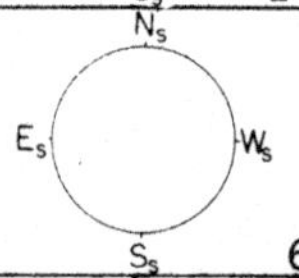	24 6
7			14	0.8	3-3+3+4- 3o3+2o1+	120 201	2 292 3	27h 31H 29f -	501 502 512 515			S27 N10	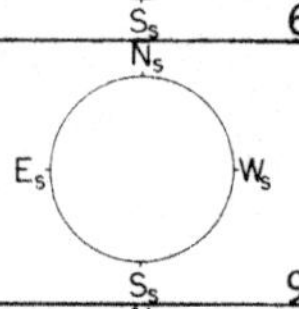	24 9
8			14	0.8	2-3-3+3o 4-3+3o2o	131 201	2 284 3	29H 28f 30c 29f	500 502 511 514		1 1		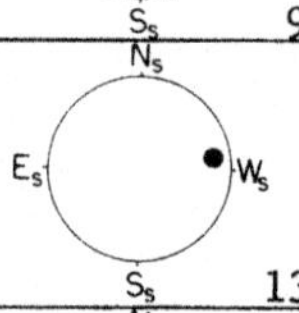	24 13.5
9			18	0.9	2-3+4+3- 3o4-3+3+	129 207	4 (305) 2	30H 31G 28f 28f	502 503 514 522	Har	1 3	N13	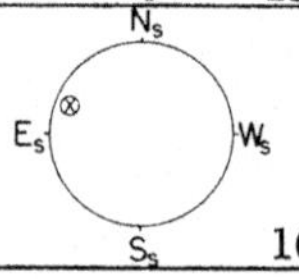	24 16.5

June 30 Moderate-to-severe magnetic disturbance continues with resurgence at 0801 UT - one of five disturbed days. Outstanding aurora. One of five disturbed World Blackout days. U.S.A. rocket ABME6.208; ionospheric physics; 83 km.; Wallops Island.

July 1 One of five quiet magnetic days. Solar electron corona, Climax. U.S.A. rocket ABME6.209; ionospheric physics; 98 km.; Wallops Island.

July 2 Slight-to-moderate magnetic activity - one of ten quiet days.

July 3 One of five quiet magnetic days.

July 4 Slight-to-moderate magnetic activity - one of ten quiet days.

July 5 Slight-to-moderate magnetic activity. Outstanding aurora. Bright surges at limb, 1515-1554 UT, 40N 90W; 2332-2416 UT, 20N 90E. Spectral radio emission II, 2338-2349 UT.

July 6 Slight magnetic activity - one of ten quiet days. Bright surge limb, 1050-1126 UT, 18N 90E.

July 7 Slight-to-moderate magnetic activity. Bright surge limb, 0335-0500 UT, 19N 90E. Spectral radio emission II, 0343.5-0345 UT.

July 8 Slight-to-moderate magnetic activity.

July 9 Slight-to-moderate magnetic activity. Spectral radio emission IV, 2044-2400 UT.

Hα flares

July 1 - 2, 0909E-1035, N16 E70, 2

July 5 - 2, 0000-0102, S22 W45, 1

July 8 - 2, 0819-0950, N11 W50, 12

July 9 - 2, 1930-2320D, N18 E67, 5

1959

JULY	GEO-ALERT	SWI	MAGNETIC INDICES Ap	C	STORM BEGINS	← 3 HR K_p →	R_Z / 3000 S	IONO-SPHERE INDEX	AUR-ORAL INDEX	COS-MIC RAY INDEX	TYPE IV OUT-BURST	SWF IMP ≥2 / <2	ACTIVE REGION LAT. / SUNSPOT LAT. AREA	FLARES POSITION IMP ≥2	HRS. PATROL / I_{AREA}
10 q			13	0.8		4o4-3+3- 1o2-2o2-	127 242	7 (288) 3	34c 32H - -	503 504 512 521		2 1	N18	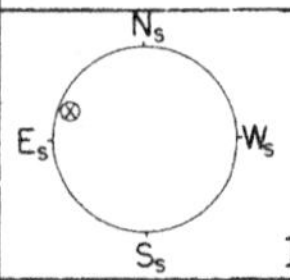	24 14
11			44	1.4	1625 (67)	4-4o3o2- 3+7-5+6+	133 201	8 C 3	24H 30G 34c 34c	502 501 506 513				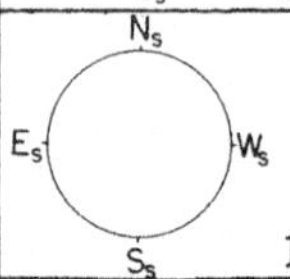	24 18
12	MAG STORM 11/ 1623		24	1.1		6+3-3-3- 4+3o1+2+	135 234	9 C 5	34c 32H 29f 32c	489 484 468 474		1 2	S14 *S14* *680*	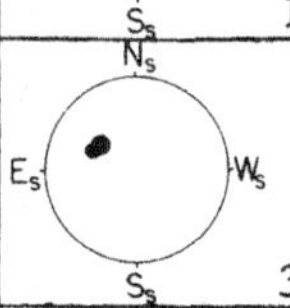	24 33
13 Q			11	0.6		2+2+3-1+ 3o3-2-3+	160 263	8 (338) 3	28G 29H - 28c	489 483 468 473	Har	2 1	N19 N15	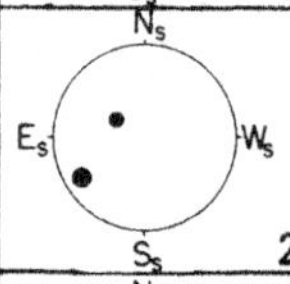	24 23.5
(14)			16	0.9		3-4-4-3o 4o3-2+2-	180 270	8 - 3	30^G_H 34G 30f -	490 484 469 474	Syd	2 1	N20 N16 *N16* *1225*	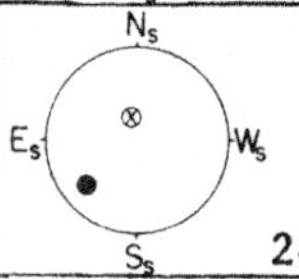	24 24
WMI (15) D	MAG STORM AUR 15/ 0802	S	236	2.0	0803 (65)	4+5o8o9- 9-9o9-9-	176 251	9 - 4	34G 38G 49D 50D	479 472 454 445		1 1	N27	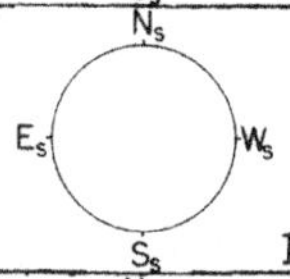	24 14
WMI (16) D		F	47	1.4		7+5+4+3o 4-3+4+5o	190 273	9 - 5	37A 35a 28c 29^c_f	470 457 411 417	Har?	1 2	N05	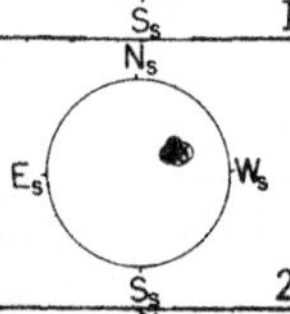	24 27
WMI 17 D			110	1.9	1624 (3) 1638 (62)	3+4+3+4- 3+8+9-8+	193 260	9 - 4	- 28f 34c 41^c_f	478 467 428 462	Har		N20 S25 N10	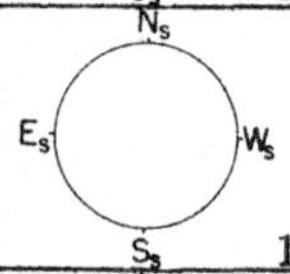	24 10
WMI 18 D	MAG STORM AUR 17/ 1638		119	2.0		8o8-8o6- 5-7o5-6o	195 234	9 - 5	34^G_H 36G 34f 42C	466 449 388 472				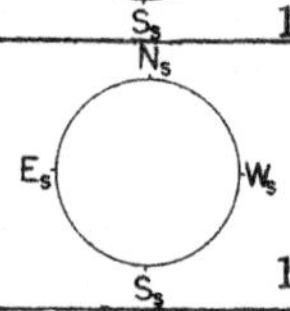	23 14
WMI 19			31	1.2		5o4+3+4o 4o5o4o3+	184 241	8 - 4	35a 33H - -	476 0 416 427			N28	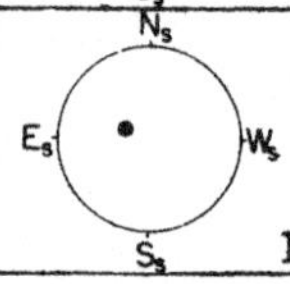	24 10.5

July 10 Slight-to-moderate magnetic activity - one of ten quiet days. Very large polar cap absorption event, 0400 UT, lasts 4 days (B,RL,K). Spectral radio emission II, 0222-0306 UT.

July 11 Moderate-to-severe magnetic storm begins with sudden commencement at 1625 UT. Forbush cosmic ray decrease (Uppsala).

July 12 Moderate-to-severe magnetic activity subsides. One of five disturbed World Blackout days.

July 13 Slight magnetic activity - one of five quiet days. Spectral radio emission IV, 1937-2005 UT.

July 14 Slight-to-moderate magnetic activity. Very large polar cap absorption event, b0700 UT, lasts 3 days (B,RL,K). Dark surges on disk, 0325-0331 and 1257-1317 UT, 15N 05E. Spectral radio emission II, 0338-0412 UT; IV, 0401-0610 UT.

July 15 Severe magnetic storm begins with sudden commencement at 0803 UT - one of five disturbed days - one of most disturbed days of IGC. Very outstanding aurora. Forbush cosmic ray decrease (Uppsala). One of five disturbed World Blackout days.

July 16 Moderate-to-severe magnetic disturbance continues - one of five disturbed days. Outstanding aurora. One of five disturbed World Blackout days. Large polar cap absorption event, b2250 UT, lasts 3 days (B,RL,K). Spectral radio emission II, 1616-1623 UT; IV,2121-2400 UT.

July 17 Severe magnetic storm begins with sudden commencements at 1624 and 1638 UT - one of five disturbed days - one of most disturbed days of IGC. Very outstanding aurora. Sea level increase in cosmic rays (Deep River). Forbush decrease (Uppsala). One of five disturbed World Blackout days. Spectral radio emission IV, 0000-0143 UT; II,0602-0606 UT.

July 18 Severe magnetic storm continues - one of five disturbed days - one of most disturbed days of IGC. Outstanding aurora. One of five disturbed World Blackout days.

July 19 Moderate magnetic activity continues.

Hα flares

July 10 - 3+, N20 E60, 0206-1000D, 10

July 12 - 2, 1410E-1515, N16 E30, 1; 2+, 2152-0020, N19 E24, 6

July 13 - 3, 0255E-0605D, N15 E18, 1; 2, 1920-2210, S23 E48, 6

July 14 - 3+, 0325E-0901, N17 E04, 18; 3, 1400-1730, S25 E37, 14

July 16 - 2, 0636-0829, N16 W20, 8; 2, 1336-1519, N21 W25, 9;
3, 1552-1645, N14 W27, 8; 3+, 2114-0030, N16 W31, 4

July 19 - 2, 0032-0220D, N12 E16, 4

1959

JULY	GEO-ALERT	SWI	MAGNETIC INDICES Ap	C	STORM BEGINS	R_z 3000 S	IONO-SPHERE INDEX	AUR-ORAL INDEX	COS-MIC RAY INDEX	TYPE IV OUT-BURST	SWF IMP ≥ 2 < 2	ACTIVE REGION LAT. SUNSPOT LAT. AREA	FLARES POSITION IMP ≥2	HRS. PATROL I_{AREA}
WMI 20			18 3o3-3+3+ (3 HR K_p)	1.0	3-4-3o4+	160 214	6 - 2	- - - -	483 0 435 440			N11	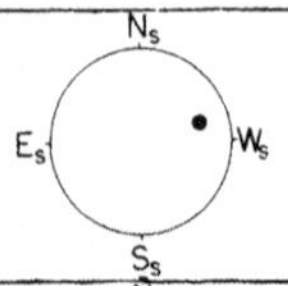	24 9.5
WMI 21			16 4-3o3-2+	0.9	4-3o3+3o	132 195	4 - 2	- 27f 30c -	485 479 439 443		2	N27	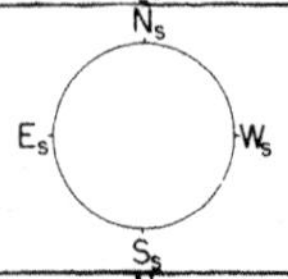	23 7
WMI 22			13 3o2o3o2o	0.8	3-3-2o4+	94 181	2 - 2	28H 31G 41c 29c	488 482 446 452				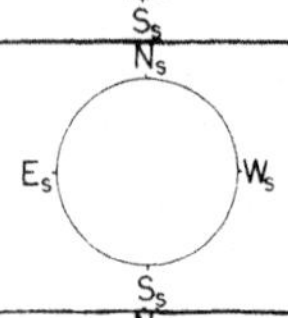	24 5
WMI 23 q			13 4o3+3o2+	0.7	3-1o2o3-	113 178	2 - 1	27G 28f - -	491 486 455 458				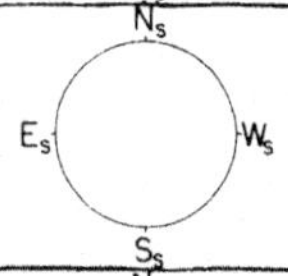	22
WMI 24			28 3o2-3-5o	1.2	3+4-5+5o	108 184	4 (301) 2	19a 28f 28f 30^c_f	491 489 462 466			S06	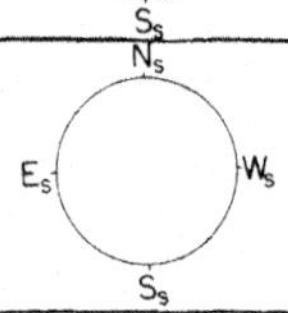	22 5
25 D			38 4+4+4+5o	1.2	4-5-5+5-	118 182	5 - 3	31H 30G 29^c_f 30c	494 492 465 469		1 2	N18 S16	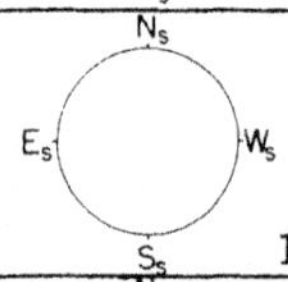	23 13
26			33 4o4+4o4o	1.1	0243 (2) 4o5-5o5-	134 189	4 (287) 4	32H 30G 29^c_f 30f	495 490 469 492		1 1		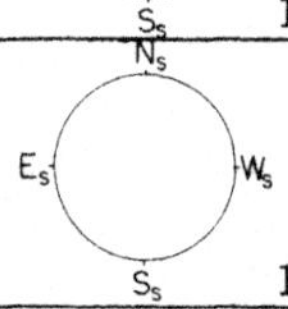	23 11
27			25 5-3+4+4-	1.1	3-3o4+4o	156 196	4 - 3	31H 31^G_H 29f 32c	497 489 474 477		1 1		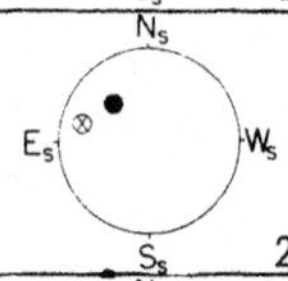	24 22.5
28			13 4-2+2+3-	0.7	3o3-2+3o	181 210	2 - 2	30H 31G 28f -	496 492 478 478			S10 S15	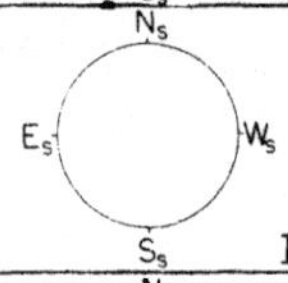	24 10
29 Q			10 2+4-3-2o	0.5	2+1+2-2-	182 208	2 (286) 1	29H 25G 27f -	498 494 481 485		2 1	N17 N25 N14	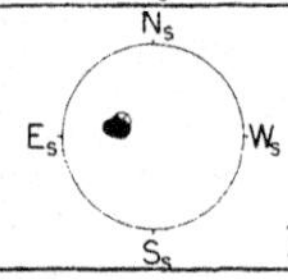	23 16

July 20 Slight-to-moderate magnetic activity. Solar electron corona, Climax.

July 21 Slight-to-moderate magnetic activity. Solar electron corona, Climax. U.S.A. rocket NN3.23F; UV and xrays; 203 km.; 1400 UT; White Sands.

July 22 Slight-to-moderate magnetic activity. Very outstanding aurora. Solar electron corona, Climax.

July 23 Slight-to-moderate magnetic activity - one of ten quiet days. Solar electron corona, Climax.

July 24 Moderate magnetic activity. Bright surges at limb, 1006-1120, 1140-1315, 1630-1700 and 1737-1850 UT, 10N 90E. Sudden disappearance of filament 2140-2228 UT, 23N 90W. U.S.A. rocket NN8.67FC; solar physics; 227 km.; 1734 UT, Point Arguello.

July 25 Moderate magnetic disturbance - one of five disturbed days.

July 26 Moderate magnetic disturbance with sudden commencement at 0243 UT. Bright surge limb, 1525-1613 UT, 20N 90E. Solar electron corona, Climax.

July 27 Moderate magnetic disturbance. Spectral radio emission II, 2117.9-2126 UT.

July 28 Slight-to-moderate magnetic activity.

July 29 Slight magnetic activity - one of five quiet days.

Hα flares

July 20 - 2, 1524-1715, N16 W44, 9

July 27 - 2, 1224-1258, N14 E50, 9; 3, 2050-2250, N27 E26, 6

July 29 - 2, 0709-0812, N11 E29, 8; 2, 1202-1408D, N11 E26, 8;
2, 2020-2234D, N10 E21, 3; 2+, 2117-2243, N15 E22, 3

1959

JULY	GEO-ALERT	SWI	MAGNETIC INDICES Ap C STORM BEGINS ← 3 HR K_p →	R_Z 3000S	IONO-SPHERE INDEX	AUR-ORAL INDEX	COS-MIC RAY INDEX	TYPE IV OUT-BURST	SWF IMP ≥2 <2	ACTIVE REGION LAT. SUNSPOT LAT. AREA	FLARES POSITION IMP ≥2	HRS. PATROL I_{AREA}
30 Q			4 0.1 2-1o0o1- 1-1-2-1o	193 208	2 (287) 0	20a 27f 27f -	499 496 483 491		2	N27	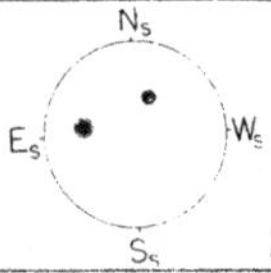	24 20
31			15 0.9 2-4-3-3+ 3o2o3o4o	190 211	3 (268) 2	34G 32G 27f 27f	502 497 484 494			N 12 N 06 N 37 N 13 N 09 N 12 1475	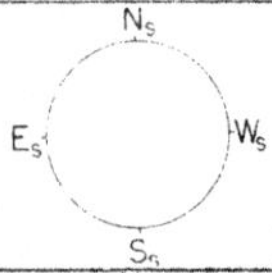	24 15
AUG 1			20 1.0 3+3+4+4- 4-4-3+2+	194 221	4 (263) 3	32H 30d 30d 28d	499 495 480 488		2	N13	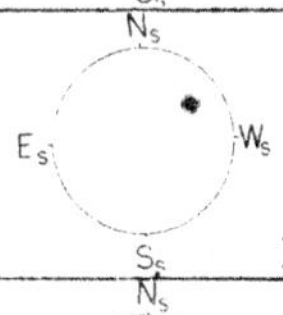	24 11.5
2			16 0.9 4o4o4-1+ 2-3-4o2-	210 234	4 270 3	35H 35H 36D 28d	498 493 479 483		1		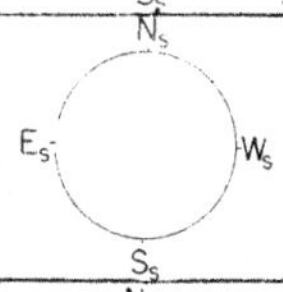	24 5
3			20 1.0 3o3o5o3o 4-4-3o1+	213 251	4 303 2	29H 35a 30c 28c	499 493 477 481		1 3	N 07 N 04 S 13 S 22 N04 620	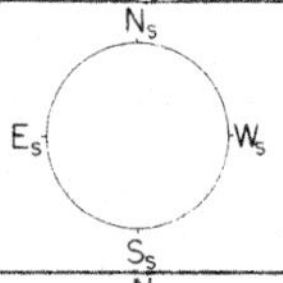	24 15
4			19 1.0 2-3-2-5- 5-4-3-3+	225 245	3 (301) 2	28G 31H 30d -	498 491 479 479		1 1		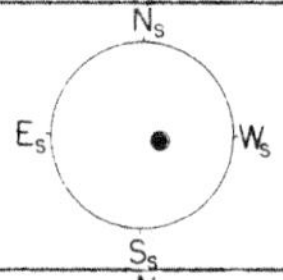	24 9.5
5 q			9 0.6 2o1+2o1+ 3-3+3-2o	212 234	2 (305) 2	21G 27d 29f 28^c_d	496 492 477 479		1	N15	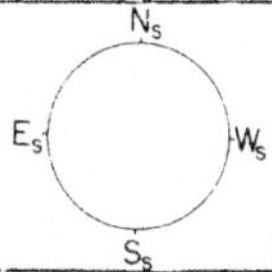	24 15
6			22 1.1 1o3o3-3+ 4o4+5o4-	207 229	3 (284) 2	28G 29f $30^{d}_{e}f$ 30^c_f	0 492 478 483		1 5	N11	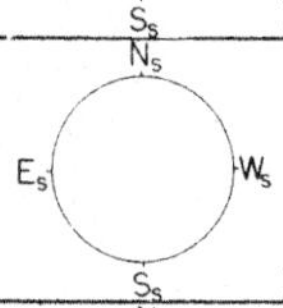	24 17
7			14 1.0 3+3o3-2+ 3-4-3o2+	179 218	2 - 2	29H 28G 28d 28d	495 490 477 486		1 1	N18 N03	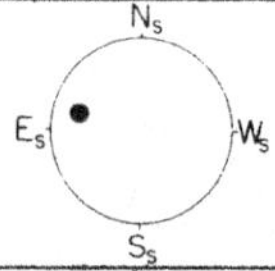	24 19.5
8			14 0.8 2+2-3-4o 3+3o1o4-	175 207	2 (275) 2	20a 27^G_H 28c -	495 490 479 484		1	S14 S17	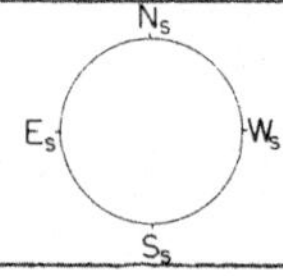	24 17

July 30 One of five quiet magnetic days - one of quietest days in IGC.

July 31 Slight-to-moderate magnetic activity.

Aug. 1 Slight-to-moderate magnetic activity. Spectral radio emission II, 1754-1810 UT.

Aug. 2 Slight-to-moderate magnetic activity. Outstanding aurora.

Aug. 3 Moderate magnetic activity. Outstanding aurora.

Aug. 4 Moderate magnetic activity

Aug. 5 Slight magnetic activity - one of ten quiet days. Sudden disappearance of filament 1808-1920 UT, 17S 90E.

Aug. 6 Moderate magnetic activity.

Aug. 7 Slight-to-moderate magnetic activity. U.S.A. rocket NN8.68FC; solar physics; 225 km.; 1805 UT; Point Arguello. U.S.A. launched 1959 delta (Explorer VI); corpuscular radiation, magnetic field, cloud cover and micrometeorites; expected life over one year.

Aug. 8 Slight-to-moderate magnetic activity.

Hα flares

July 30 - 2, 0000-0117, N11 E35, 3; 2, 1648-1850D, N29 W11, 3

Aug. 1 - 2, 1313-1450, N27 W37, 7

Aug. 4 - 2, 1028-1110, N03 W12, 5

Aug. 7 - 2, 1551-1636, N15 E45, 6

1959

AUGUST	GEO-ALERT	SWI	MAGNETIC INDICES Ap	C	STORM BEGINS	← 3 HR K_p →	R_Z 3000 S	IONO-SPHERE INDEX	AUR-ORAL INDEX	COS-MIC RAY INDEX	TYPE IV OUT-BURST	SWF IMP ≥2 <2	ACTIVE REGION LAT. SUNSPOT LAT. AREA	FLARES POSITION IMP ≥2	HRS. PATROL I_{AREA}
9			22	1.1		4o1+3+3+ 4o4+5-3-	170 198	3 (312) 2	26H 29f 30^{c}_{f} 30c	497 492 481 484			N17 S09	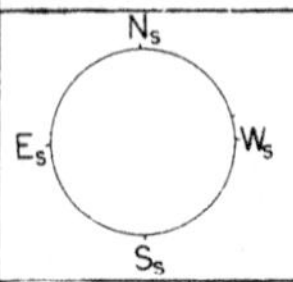	24 6
10			14	0.8		3o4+2-1+ 3o3-3+2+	155 199	2 - 3	30H 28G 29f -	496 492 480 482			N27 N21 N15	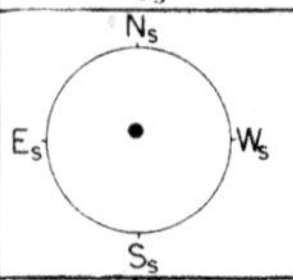	24 6.5
(11) q			8	0.4		3-3o2o1+ 2o3-1-2-	180 200	1 (281) 2	32H 32H 29f -	498 494 483 486			N15	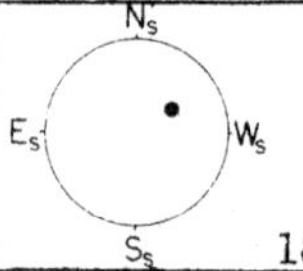	24 18.5
(12) Q			5	0.2		2-1+2-1o 2-1+1-2o	160 189	0 303 1	32H 28G - 28c	497 494 484 489			S20	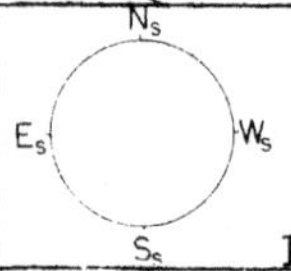	24 11
(13) Q			6	0.3		2-2-1+2- 1+1+2o2-	125 189	0 (275) 1	24H 27H - 28c	498 496 486 492				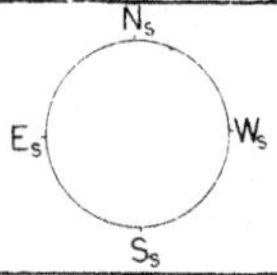	23 5
14 Q			6	0.3		2-2-1+1+ 2-2-2o1+	139 189	1 - 1	28^{c}_{H} 27f 32c 34c	499 495 487 491		1	S07	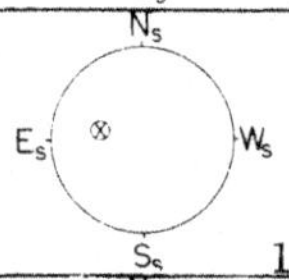	24 11.5
15			27	1.2		3-2-1+3+ 4o5o4+6-	144 194	1 (289) 3	- 29f - 35a	500 498 487 489			N18 N21 N23 N04	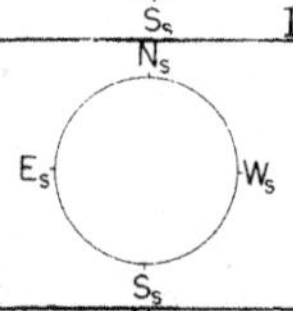	24 4
16 D	MAG STORM 15/ 1400		130	2.0	0404 (48)	4+4+7-7+ 8+8o7-7+	157 197	6 C 5	35a 38G 40D 39D	496 494 481 481		1	N13 N16 *N12* *730*	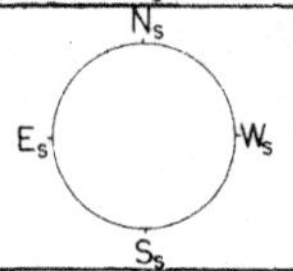	24 2
17 D			114	1.8		8-8+7-6+ 5+7o5o5+	166 217	7 257 5	37H 37^{G}_{H} 37D 33C	498 496 491 488		4 2		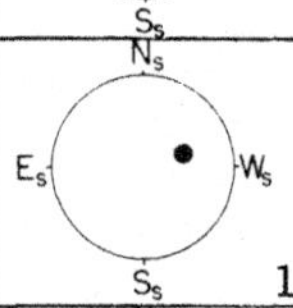	23 15.5
18			28	1.2		3o4+5+3o 4+4o5-3-	174 226	5 - 4	21G 33G 18c 30c	499 496 490 492	Har	4 1	N20 N04	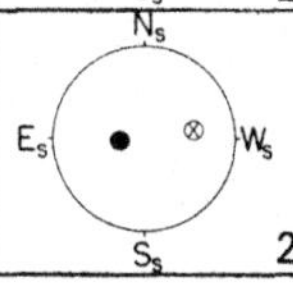	23 24.5

Aug. 9 Moderate magnetic activity. Dark surges on disk, 0920-0948 and 1428-1500 UT, 15N 15E. Sudden disappearance of filament 2318-2345 UT, 01N 90W.

Aug.10 Slight-to-moderate magnetic activity. Sudden disappearance of filament, 0555-0835 UT, 02S 05E; 1516-1630 UT 45N 90W; 2204-2218 UT, 11N 03E. Solar electron corona, Climax.

Aug.11 Slight magnetic activity - one of ten quiet days. Solar electron corona, Climax.

Aug.12 One of five quiet magnetic days.

Aug.13 One of five quiet magnetic days.

Aug.14 One of five quiet magnetic days. U.S.A. rocket NN8.69FC; solar physics; 217 km.; 1700 UT; Point Arguello.

Aug.15 Moderate-to-severe magnetic activity begins. Outstanding aurora.

Aug.16 Severe magnetic storm with sudden commencement at 0404 UT - one of five disturbed days - one of most disturbed days of IGC. Very outstanding aurora. One of five disturbed World Blackout days. Solar electron corona, Climax.

Aug.17 Severe magnetic storm continues - one of five disturbed days - one of most disturbed days of IGC. Outstanding aurora. One of five disturbed World Blackout days. Spectral radio emission II, 0334-0341 UT. Solar electron corona, Climax. U.S.A. rocket NASA3.13; upper atmos. physics; 236 km; 0918 UT; Wallops Island.

Aug.18 Moderate magnetic activity. One of the five disturbed World Blackout days. Spectral radio emission IV, 1717-1810, 1827-1835, 1838-1853, 1917-1922, 1959-2007, 2022-2027, 2029-2127, 2220-2225 and 2239-2243 UT.

Hα flares

Aug.10 - 2, 2200-2333, N12 E02, 3

Aug.11 - 2, 1203-1232, N20 W24, 7

Aug.14 - 2, 0040-0326, N11 E27, 5

Aug.17 - 2, 2046-2113, N14 W27, 4

Aug.18 - 3, 1014-1350, N12 W33, 18; 2+, 1654-1822, N05 W16, 5

1959

AUGUST	GEO-ALERT	SWI	MAGNETIC INDICES Ap C STORM BEGINS ← 3 HR K_p →	R_z 3000S	IONO-SPHERE INDEX	AUR-ORAL INDEX	COS-MIC RAY INDEX	TYPE IV OUT-BURST	SWF IMP ≥2 <2	ACTIVE REGION LAT. SUNSPOT LAT. AREA	FLARES POSITION IMP ≥2	HRS. PATROL I AREA
19			21 1.1 3+5-4+3o 2o3o4o3o	182 214	6 (268) 4	30H 32H 32c 32c	497 495 483 488		 1	N22 N09 *N06* *1025*	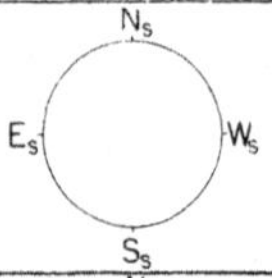	22 9
20 D	MAG STORM 20/ 0410		38 1.4 0412 (58) 3-6-6-5o 4o2+5o4o	180 225	5 (289) 3	24H 32G 29f 29c	489 481 459 462		1 1	N06	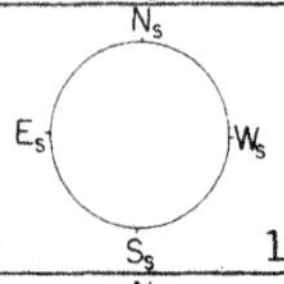	24 13
21 D			34 1.3 5-4-4o5o 3+4o5+5-	200 234	5 299 3	28A 30f 29f 30^c_c	493 486 469 470		 1	N11 S23	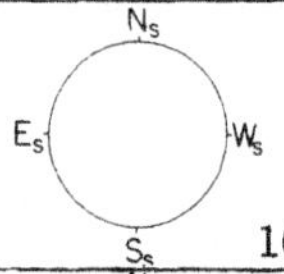	24 16
22			27 1.1 5+4-4+3o 3o4-5o4o	204 247	4 - 2	29H 30f 29f 28C	493 489 475 475		2 1	N28 N10	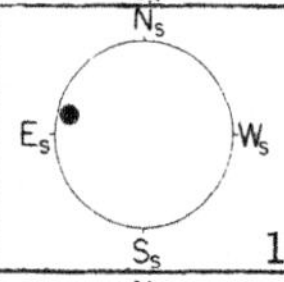	24 11.5
23 D			28 1.2 4-3+4-4- 5-5-4+4+	205 263	3 348< 2	30H 31G 33E 27C	495 490 475 479		 1		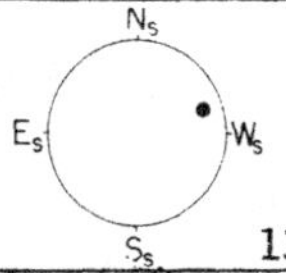	24 13.5
24			20 1.1 4-4-5o3+ 2+3o2o3o	217 255	2 286 1	29G 33G 28^c_d 28c	498 492 479 483		1		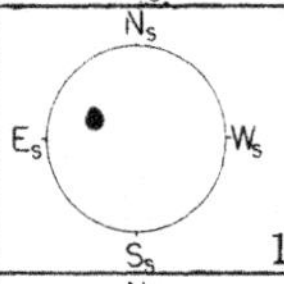	22 12
25			15 0.9 3o3+3o3+ 3o3+1o3o	212 258	2 (331) 1	28h 29G 29f 29d	498 495 486 485		1 2	N17	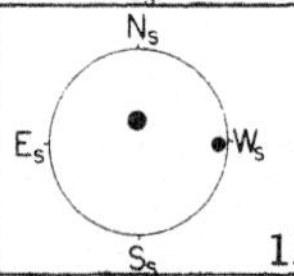	23 15
26 q			8 0.4 3o2-3o2- 1o1+2o2-	220 255	1 - 1	28c 29f 27^c_f 28^c_d	500 497 483 483		1	N17 *N15* *1650*	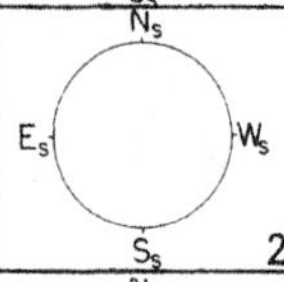	21 24
27 Q			5 0.2 3-1o2o1+ 1o0+0+0+	231 276	1 (342) 0	- 22^G_H - 23c	502 499 487 487	Har		N13 N26	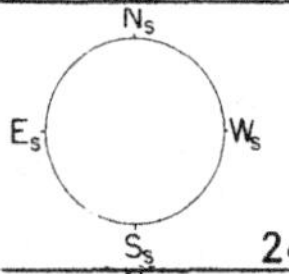	24 24
28 Q			3 0.1 0o1-1+1o 1-1-0+1+	274 303	0 (365)< 0	18c 27H 33E 17c	503 500 489 490	Har	1	N10 S11 N09 *S10* *2050*	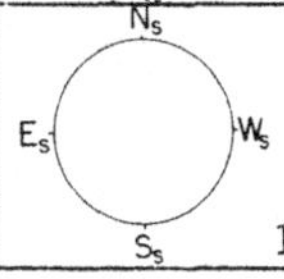	24 15

Aug.19 Moderate magnetic activity. One of the five disturbed World Blackout days. Bright surges at limb, 0424-0446, 0630-0652 and 0823-0835 UT, 18N 90E. U.K. rocket Skylark - 17; ion probe and atmospheric pressure; 0050 UT; Woomera.

Aug.20 Moderate-to-severe magnetic storm with sudden commencement at 0412 UT - one of five disturbed days.

Aug.21 Moderate magnetic disturbance - one of five disturbed days. One of the five disturbed World Blackout days.

Aug.22 Moderate magnetic activity. Sudden disappearance of filament 1045-1200 UT, 32S 90W. Solar electron corona, Climax. U.S.A. rocket AA3.101C; atmospheric composition; 230 km.; 0045 UT; White Sands.

Aug.23 Moderate magnetic disturbance - one of five disturbed days. Sudden disappearance of filament 1155-1400 UT, 11N 90W.

Aug.24 Moderate magnetic activity. Solar electron corona, Climax.

Aug.25 Slight magnetic activity. Bright surge on disk and at limb, 0504-0517 UT, 06S 65W. Sudden disappearance of filament 0620-0740 UT, 30S 90W. U.S.A. rocket NN8.71FC; solar physics; 0047 UT; Point Arguello.

Aug.26 Slight magnetic activity - one of ten quiet days. Solar electron corona, Climax.

Aug.27 One of five quiet magnetic days. Spectral radio emission IV, 0000-0113 UT.

Aug.28 One of five quiet magnetic days - one of quietest days in IGC. Spectral radio emission II, 0039-0127 UT; IV, 0055-0330 UT. Solar electron corona, Climax. U.S.A.rocket NN8.72FC; solar physics; 214 km.; 1710 UT, Point Arguello. U.S.A. rocket AA3.122C; ionospheric physics; 235 km.; 1845 UT; White Sands.

Hα flares

Aug.22 - 2, 1310-1345, N16 E60, 7

Aug.23 - 2, 0750-0835, N19 W54, 1

Aug.24 - 2, 0122-0215D, N20 E30, 3; 2+, 2233-2332, N17 E28, 2

Aug.25 - 2, 0457-0522, N00 W64, 2; 2, 0623-0728, N20 E01, 10

1959

AUGUST	GEO-ALERT	SWI	MAGNETIC INDICES Ap	C	STORM BEGINS	3 HR Kp	R_Z 3000 S	IONO-SPHERE INDEX	AUR-ORAL INDEX	COS-MIC RAY INDEX	TYPE IV OUT-BURST	SWF IMP ≥2 <2	ACTIVE REGION LAT. SUNSPOT LAT. AREA	FLARES POSITION IMP ≥2	HRS. PATROL I AREA
29			13	0.7		3o2-1+3- 3-4o3-3-	301 319	1 338 0	29G 21G 29d 28 c/d	501 499 490 490		4	N05 N15	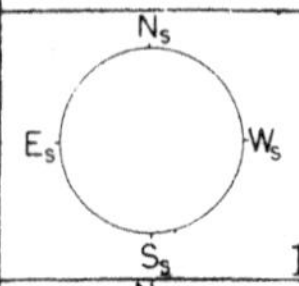	24 16
30 q			9	0.6		3o2o2o2o 2+3o2+2-	292 324	1 C 1	17c 27G 32f 18c	497 492 484 488		2	N10 S15	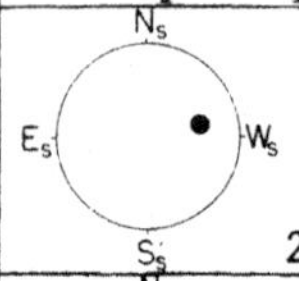	24 21.5
31 q			9	0.6		2-2+2+2+ 3o2+2+2+	284 308	1 342 2	26H 32f 29f 34c	497 495 488 489	Har	2 2	S04 S07 N09 600	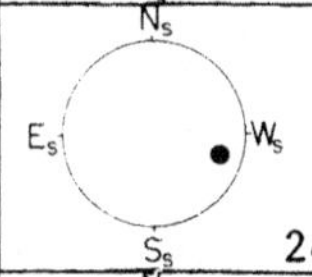	24 24.5
SEP 1			20	1.1		3+3o3+4- 3o4o4+3-	290 300	4 (344) 2	31G 31H 34D 34D	499 496 488 489	Har	2 2	N10	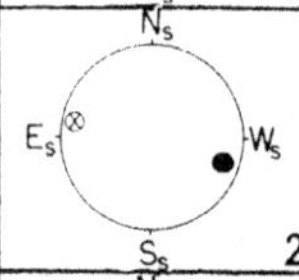	24 28.5
2	MAG STORM 02/ 0110		34	1.2		4+5+5o5- 3-3+4o5-	256 296	3 346 3	33G 36f 32c 36D	500 497 490 492		2 2	N11	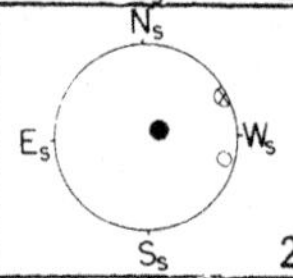	24 23.5
3			33	1.4	1417 (35) 2159 (52)	3o3-3+4- 3o4-4o7o	202 270	3 (328) 2	29H 33G 34D 36 C/D	498 495 489 489		2	S 20 N 17 N 17 720	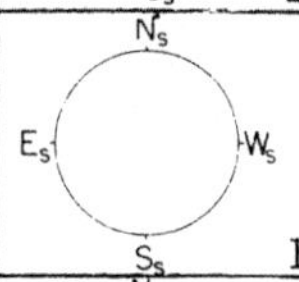	23 13
4 D	MAG STORM AUR 03/ 2159	S	103	1.8		7-7+7o6+ 6o7o5+6-	161 247	6 307 4	39 G/H 39 e/G 37E 37C	0 483 470 464			S20 N03	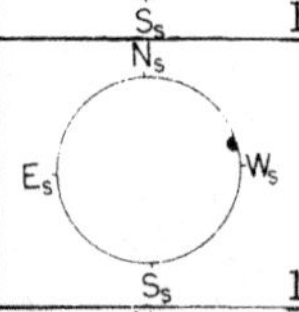	23 11.5
5		F	34	1.3		5o4-4o4o 4o4-4+6-	148 230	5 287 4	37C 31H 34 D/E 34 c/D	486 478 467 460		1	N12 S11	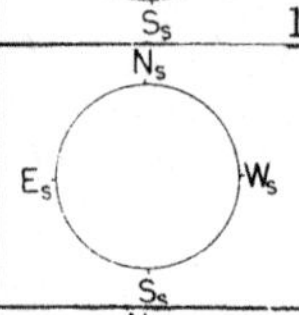	23 3
6			18	0.9		6-4-4-3- 2+2+1o1-	152 202	1 286 2	33 C/H 30 G/H 36D 36D	491 483 470 466			N21	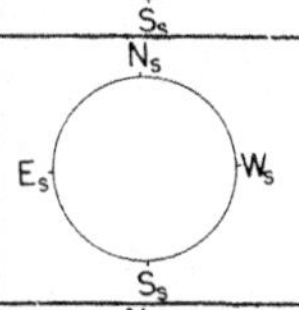	24 4
7 Q			7	0.4		1+1-1+2o 2+3+2-2-	135 205	1 291 1	32H 32H 20c 35D	493 483 473 473				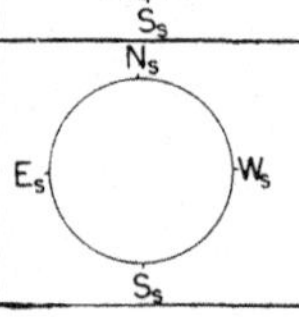	24 4

Aug.29 Slight-to-moderate magnetic activity. Solar electron corona, Climax.

Aug.30 Slight magnetic activity - one of ten quiet days.

Aug.31 Slight magnetic activity - one of ten quiet days. Spectral radio emission IV, 1859-1916 UT. Solar electron corona, Climax.

Sep. 1 Slight-to-moderate magnetic activity. Sudden disappearance of filament 1850-1935 UT, 10N 70E. Spectral radio emission IV,1914-1917, 1920-1922, 1927-1939, 1940-1941 and 1945-1950 UT; II, 1939.3-1945 UT. Solar electron corona, Climax. U.S.A. rocket,NN8.73FC; solar physics; 0053 UT; Point Arguello. U.S.A. rocket NN8.75 FC; solar physics; 210 km.; 1800 UT; Point Arguello.

Sep. 2 Moderate magnetic disturbance. Outstanding aurora. Very small polar cap absorption event, 0400 UT, lasts 2 days (B). Spectral radio emission II, 1608.1-1615 UT.

Sep. 3 Moderate-to-severe magnetic storm with sudden commencements at 1417 and 2159 UT. Outstanding aurora. Bright surges at limb, 0425-0455 and 1734-1901 UT, 25N 90W. Spectral radio emission II, 0424.5-0438 UT. Solar electron corona, Climax.

Sep. 4 Moderate-to-severe magnetic storm - one of five disturbed days - one of most disturbed days of IGC. Outstanding aurora. One of the five disturbed World Blackout days. Sudden disappearance of filament 0710-0800 UT, 15N 85W. Solar electron corona, Climax.

Sep. 5 Moderate-to-severe magnetic activity. Outstanding aurora. Spectral radio emission II, 1602.6-1610 UT. Solar electron corona, Climax.

Sep. 6 Moderate magnetic activity. Outstanding aurora.

Sep. 7 One of five quiet magnetic days. Outstanding aurora. Solar electron corona, Climax.

Hα flares

Aug.30 - 2, 1535-1700, N13 W36, 5

Aug.31 - 2+, 2222-2344, S08 W46, 4

Sep. 1 - 2, 1359-1450D, S12 W51, 5; 2, 1648-1832, S12 W52, 8; 2+, 1933-2216, N12 E60, 3

Sep. 2 - 2, 0310E-0512, S12 W63, 1; 2+, 0720-0954, N10 W10, 13; 2, 1602-1645, N25 W77, 8

Sep. 4 - 2, 0730-0820, N15 W80, 1

1959

SEPTEMBER	GEO-ALERT	SWI	MAGNETIC INDICES Ap C STORM BEGINS ← 3 HR K_p →	R_z 3000 S	IONO-SPHERE INDEX	AUR-ORAL INDEX	COS-MIC RAY INDEX	TYPE IV OUT-BURST	SWF IMP ≥ 2 < 2	ACTIVE REGION LAT. SUNSPOT LAT AREA	FLARES POSITION IMP ≥2	HRS. PATROL I_{AREA}
8 Q			10 0.6 2-2o3-3+ 2+3o1+2+	136 204	1 (287) 1	26^{C}_{H} 29^{C}_{G} 36D 36D	495 487 480 478		.	N14 N28	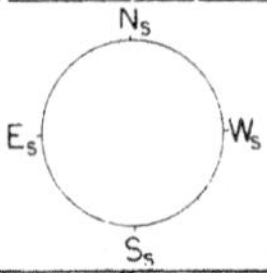	24 8
9 Q			6 0.2 2-1o1o2+ 1o2-1+2o	157 215	1 276 0	28c 30H 32D 33C	495 489 481 485		1 1		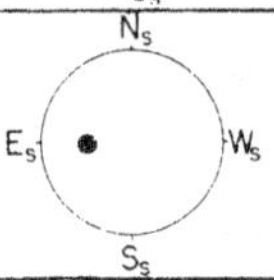	24 6.5
10 Q			8 0.4 2-1+1o1+ 2-2o2-4-	141 208	1 258 1	22H 26G 29E 31C	496 491 484 489			S18 N13	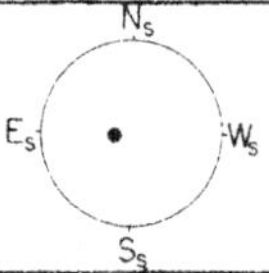	24 7.5
11 q			15 0.9 2o3+3o4- 4-2+3o3-	155 207	1 275 1	29C 31^{G}_{H} 32D 34c	497 492 489 495			N17 N06 S17 N12 S18 630 N06 1025	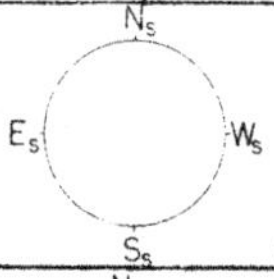	24 6
12 q			14 0.8 5-3o3-3- 2+3o2o1+	170 200	1 272 1	31H 29^{f}_{G} 30f 29C	499 493 491 496				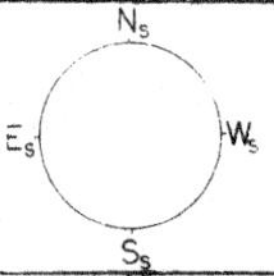	24 1
RWI 13 q			14 0.7 3+1o2o1+ 2-1o2+5+	148 190	1 (270) 0	28H 31H 32C 33C	498 494 492 497		1	N16	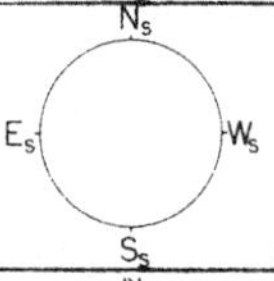	23 8
RWI 14			18 1.1 4-3+4-3+ 4-3+2o3+	151 200	2 284 2	30^{C}_{H} 29^{AG}_{e} 33D 33D	498 495 495 498		1		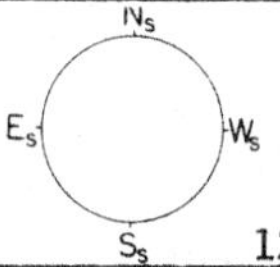	24 12
RWI (15)			18 1.1 2+2-3+3- 4-3+5-4-	161 189	2 289 2	28C 29^{f}_{G} 32C 33D	498 494 495 497			S17	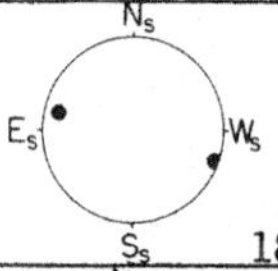	24 18.5
RWI (16) q			16 0.9 4+5-2+3o 2+3-3-2+	130 179	2 264 2	33G 29G 30E 29D	500 495 497 500		1	N05	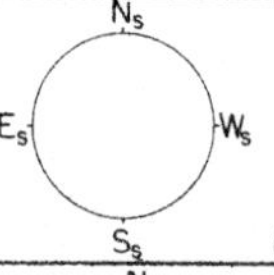	24 6
RWI (17)			23 1.2 2+3-2o3o 3+4o5o5o	87 172	1 269 2	28G 27H 34E 33D	500 496 501 503				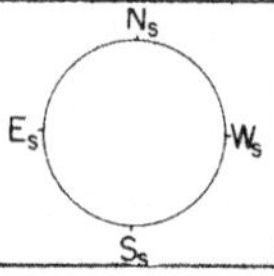	23 4

Sep. 8 Slight magnetic activity - one of five quiet days. Outstanding aurora. Solar electron corona, Climax.

Sep. 9 One of five quiet magnetic days. Forbush cosmic ray decrease (Uppsala). Solar electron corona, Climax.

Sep.10 Slight magnetic activity - one of five quiet days.

Sep.11 Slight-to-moderate magnetic activity - one of ten quiet days. Solar electron corona, Climax. U.S.A. rocket NASA4.08; ionospheric physics; 225 km.; 1912 UT; Ft. Churchill.

Sep.12 Slight-to-moderate magnetic activity - one of ten quiet days. U.S.S.R. launched space probe Lunik II; temperatures, pressures, heavy primary cosmic rays, magnetic fields, micrometeorites; lunar impact.

Sep.13 Slight-to-moderate magnetic activity - one of ten quiet days.

Sep.14 Slight-to-moderate magnetic activity. U.S.A. rocket NASA4.07; ionospheric physics; 219 km.; 1728 UT; Ft. Churchill.

Sep.15 Moderate magnetic activity. Spectral radio emission II, 2124.3-2127 and 2135-2144 UT.

Sep.16 Moderate magnetic activity but one of ten quiet days. Bright surge limb, 1850-1940 UT, 33N 90E.

Sep.17 Moderate magnetic activity. U.K. rocket Skylark - 12; solar x-ray cameras and solar Lyman - alpha counter; 0042 UT; Woomera. Canadian rocket DRTE01; electron density measurements and cosmic ray measurements.

Hα flares

Sep. 9 - 2, 0645-0925, N05 E29, 14

Sep.10 - 2, 1225E-1230D, N04 E11, 1

Sep.15 - 2, 0558E-0602D, S17 W73, 1; 2, 0724E-0915, N14 E57, 3; 2, 0759-0902, S17 W72, 4

1959

SEPTEMBER	GEO-ALERT	SWI	MAGNETIC INDICES Ap	C	STORM BEGINS	3 HR Kp	Rz / 3000 S	IONO-SPHERE INDEX	AUR-ORAL INDEX	COS-MIC RAY INDEX	TYPE IV OUT-BURST	SWF IMP ≥2 / <2	ACTIVE REGION LAT. / SUNSPOT LAT. AREA	FLARES POSITION IMP ≥2	HRS. PATROL / I_{AREA}
RWI 18			27	1.2		3+3-3o5- 4o4-4+5o	100 173	1 280 2	25G 30f 32f 32D	501 497 500 502			S16	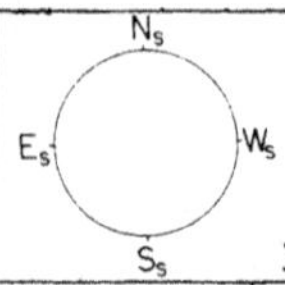	24 3
RWI 19	MAG STORM 18/ 2100		36	1.2	1052 (3) 1128 (38)	6o6o3+5o 4+2+3-3-	120 175	5 285 2	35H 32H 30E 26C	497 493 485 484			N13	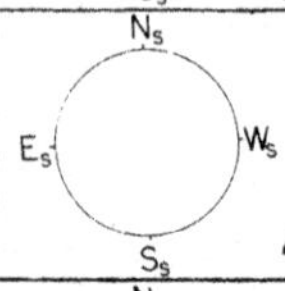	24 4
RWI 20 D		S	61	1.6	1157 (43)	4o5o5+5o 6o6-6+6-	149 180	6 294 5	32C 31G 41D 41D	496 492 487 486		1	S12	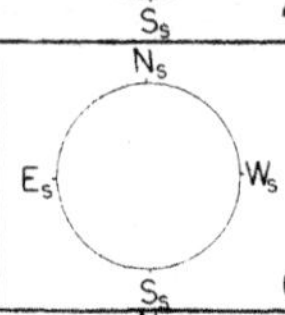	24 6
RWI 21 D		F	86	1.9		6-7-6o5- 6o7-6+5+	143 185	7 260 5	35a 34$^{f}_{H}$ 35E 40D	496 493 492 492		1	N20	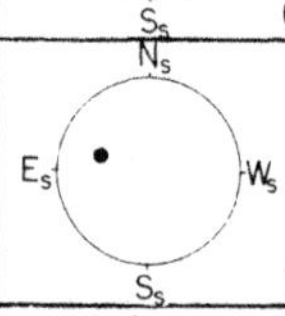	24 7.5
RWI 22 D			73	1.7	0304 (6)	6+7-7o6- 6-5o5+2+	157 187	5 253 3	35G 35G 35E 35D	498 496 496 495		1	N03 N20	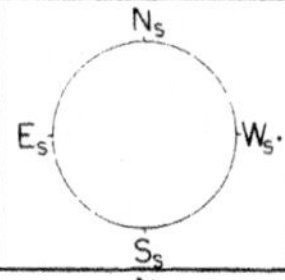	24 1
23			24	1.2		3-3-4-4- 2+5-4+5o	143 186	4 250 2	27C 31E 34c 34c	499 497 499 501		1		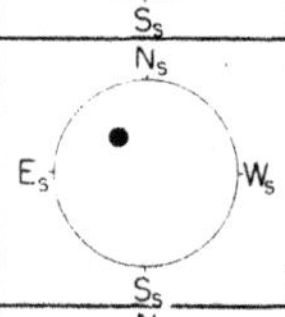	24 7.5
24			24	1.2		4o4o4-3+ 5o3o3+3o	155 177	4 247 3	32H 29$^{G}_{H}$ 34E 34$^{D}_{E}$	501 498 503 502			N27 N19 S09 N26 S12	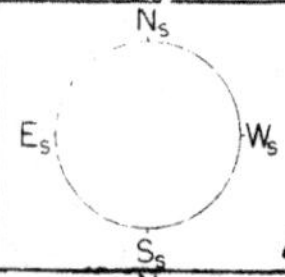	23 4
25 D	MAG STORM 25/ 0045		37	1.2		5o5-4o4+ 4o6-3+5-	132 174	5 - 3	33H 34E 38D 39D	499 497 503 500			S11	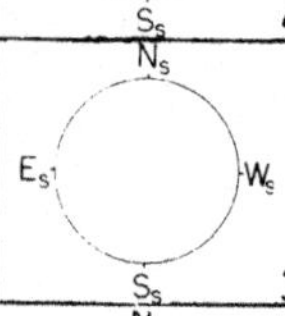	22 3
26			23	1.2		4o3o3+4- 5-3o4-4o	110 169	5 (265) 3	33H 36E 34E 33D	499 498 500 498				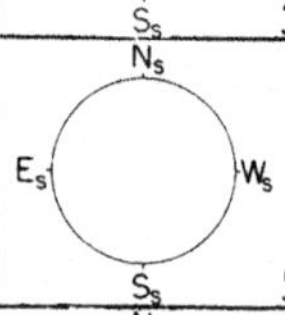	24 5
27			21	1.1		3o3+4-4o 4-4+3o3+	102 156	4 251 2	28C 30$^{EG}_{f}$ 35E 33D	499 498 498 497			S05	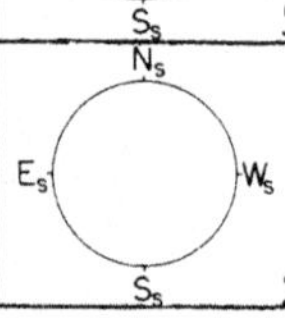	23 2

Sep.18 Moderate magnetic activity. U.S.A. launched 1959 eta (Vanguard III); temperatures, magnetic field, solar x-radiation and micrometeorites; estimated life 30-40 years.

Sep.19 Moderate-to-severe magnetic storm with sudden commencements at 1052 and 1128 UT. Outstanding aurora. Sudden disappearance of filament 0715-0850 UT, 25S 17E.

Sep.20 Moderate-to-severe magnetic storm continues with resurgence at 1157 UT - one of five disturbed days. Very outstanding aurora. One of the five disturbed World Blackout days.

Sep.21 Moderate-to-severe magnetic storm continues - one of five disturbed days. Very outstanding aurora. One of the five disturbed World Blackout days.

Sep.22 Moderate-to severe magnetic storm continues with resurgence at 0304 UT - one of five disturbed days. Outstanding aurora. One of the five disturbed World Blackout days.

Sep.23 Moderate magnetic disturbance. Sudden disappearance of filament 0821-1005 UT, 26S 90E. Solar electron corona, Climax.

Sep.24 Moderate magnetic activity. U.K. rocket Skylark - 15; ion mass spectrometer and sporadic-E detector; 0059 UT; Woomera.

Sep.25 Moderate-to-severe magnetic disturbance - one of five disturbed days. Outstanding aurora. One of the five disturbed World Blackout days.

Sep.26 Moderate magnetic disturbance. Outstanding aurora. Sudden disappearance of filament, 2205-2330 UT, 17N 90E; 2240-2407 UT, 25N 90W.

Sep.27 Slight-to-moderate magnetic activity. Outstanding aurora. Sudden disappearance of filament, 2325-2400 UT, 45S 15W; 1200-1250 UT, 40N 14E.

Hα flares

Sep.21 - 2, 2026-2222D, N25 E33, 3

Sep.23 - 2, 0419-0504, N29 E20, 2

1959

SEPTEMBER	GEO-ALERT	SWI	MAGNETIC INDICES Ap	C	STORM BEGINS	← 3 HR K_p →	R_Z / 3000S	IONO-SPHERE INDEX	AUR-ORAL INDEX	COS-MIC RAY INDEX	TYPE IV OUT-BURST	SWF IMP ≥2 / <2	ACTIVE REGION LAT. / SUNSPOT LAT AREA	FLARES POSITION IMP ≥2	HRS. PATROL / I_{AREA}
28			18	0.9		4o3o4-3+ 4-3-3+3-	91 158	2 258 2	31H 31E 34E 32C	499 499 498 502				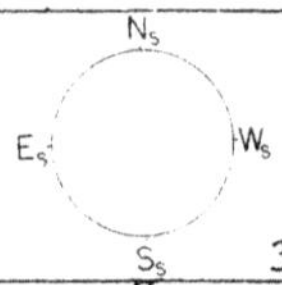	22 3
29 Q			9	0.5		3o2-2o3+ 2+1+1-3-	87 162	2 249 1	32C 23G 33E 32C	500 501 498 504			N09	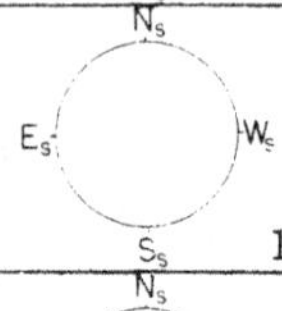	21 1
30 q			12	0.9		2o1+2-3+ 3+3o3+3-	86 159	1 266 1	33H 34E 36D 35D	501 501 501 505				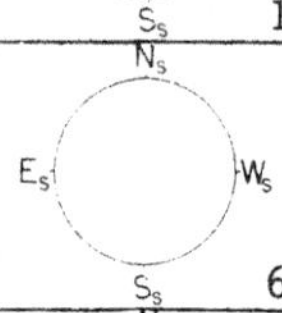	24 6
OCT (1) D	MAG STORM 01/ 0430		44	1.3		4o5+6o6- 4o4o4o5-	65 153	3 237 3	35G 39E 39E 35D	504 504 504 504			S21 S11 N07	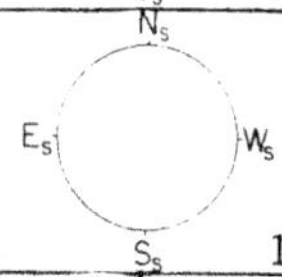	24 1
(2)			23	1.0		3-6-5o4o 2o2+2+2-	76 149	1 259 3	35D 35E 36D 34c	502 502 498 497				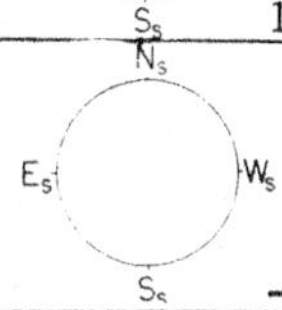	22 -
(3) D			46	1.5	1711 (3)	1o3-4o3o 3+6-7-7-	89 146	4 233 2	33G 35E 37E 36D	500 502 499 497				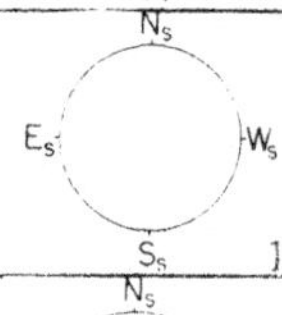	22 1
4 D	MAG STORM AUR 03/ 17XX		51	1.3	1856 (2)	5o7o5+5- 6-3o4+3+	101 148	5 187 3	36D 36E 35E 36D	501 504 500 499				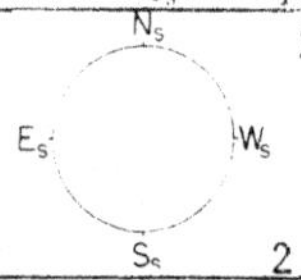	21 2
5			37	1.3	1604 (8)	5-4+3o3- 3+4+6o6-	115 159	4 204 3	32H 34E 39E 39 D/E	500 502 500 498			S10 S19	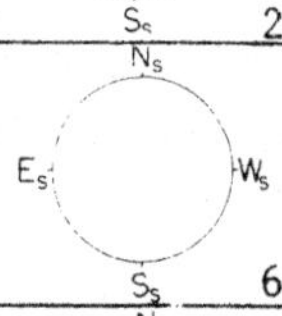	21 6
6 D			53	1.5		6-6-5+5+ 5-5o5+4+	128 173	5 208 4	35 C/H 38E 38E 38E	500 502 501 499		3	S10	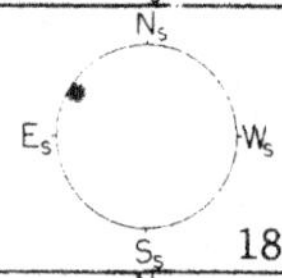	22 18.5
7			15	0.9		3+4-3-3+ 3-1+2+4o	130 170	4 224 2	27H 32 C/H 37E 35D	501 499 498 496		5	S18	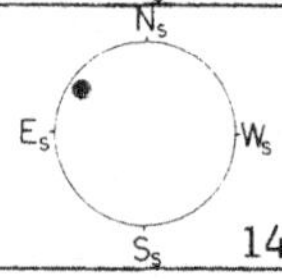	22 14.5

Sep.28 Slight-to-moderate magnetic activity.

Sep.29 Slight magnetic activity - one of five quiet days. U.S.A. rocket AA6.203C; atmospheric composition and upper atmos. physics; 124 km.; 1059 UT; Eglin Field.

Sep.30 Slight magnetic activity - one of ten quiet days. Outstanding aurora. U.S.A. rocket AA6.204C; atmospheric composition and upper atmos. physics, 122 km.; 1057 UT; Eglin Field.

Oct. 1 Moderate-to-severe magnetic disturbance - one of five disturbed days. Outstanding aurora. U.S.A. rocket AA6.210C; atmospheric composition and upper atmos. physics; 138 km.; 1049 UT; Eglin Field. U.S. Meteorological Rocket Network launched 18 rockets during October 1959.

Oct. 2 Moderate-to-severe magnetic activity. Outstanding aurora. U.S.A. rocket AA6.211C; atmospheric composition and upper atmos. physics, 143 km.; 1100 UT; Eglin Field.

Oct. 3 Moderate-to-severe magnetic storm begins with sudden commencement at 1711 UT - one of five disturbed days. Outstanding aurora. One of the five disturbed World Blackout days. U.S.A. rocket AA6.202C; atmospheric composition and upper atmos. physics; 122 km.; 1103 UT, Eglin Field.

Oct. 4 Moderate-to-severe magnetic storm continues with resurgence at 1856 UT - one of five disturbed days. Outstanding aurora. One of the five disturbed World Blackout days. U.S.S.R. launched 1959 theta (Lunik III); lunar photography; down April 1960.

Oct. 5 Moderate-to-severe magnetic disturbance continues with resurgence at 1604 UT. Outstanding aurora. One of the five disturbed World Blackout days. Solar electron corona, Climax.

Oct. 6 Moderate-to-severe magnetic disturbance continues - one of five disturbed days. Outstanding aurora. One of the five disturbed World Blackout days .

Oct. 7 Slight-to-moderate magnetic activity. Outstanding aurora.

Hα flares

Oct. 6 - 2, 1405-1448, N30 E64, 8

Oct. 7 - 2, 0502-0544, N31 E57, 3

1959

OCTOBER	GEO-ALERT	SWI	MAGNETIC INDICES Ap / C / STORM BEGINS / ← 3 HR Kp →	R_Z / 3000 S	IONOSPHERE INDEX	AURORAL INDEX	COSMIC RAY INDEX	TYPE IV OUTBURST	SWF IMP ≥2 / <2	ACTIVE REGION LAT. / SUNSPOT LAT. AREA	FLARES POSITION IMP ≥2	HRS. PATROL / I_{AREA}
8			8 0.3 3o3-2-3- 1+2-1o1-	115 164	1 220 1	33C 29^{E}_{f} 30E 28^{C}_{D}	500 500 499 497			N04 N05 N08	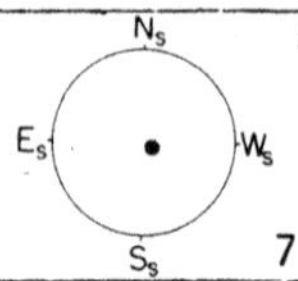	24 7.5
(9) q			6 0.3 3-1+2-2- 1-1o3-0+	103 157	1 231 1	27^{C}_{D} 24H 31^{D}_{E} 27C	502 501 502 494			N21	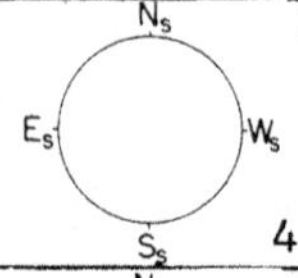	24 4
(10) Q			4 0.2 1o1o1o1o 1-2+1+1-	91 156	1 234 0	29H 23G 34E 32D	502 502 504 502		1 1		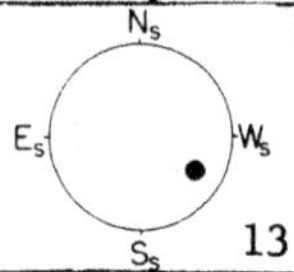	24 13.5
(11) Q			3 0.2 1o0o1o0+ 0o0o2-1+	87 150	0 244 1	28D 34E 30D 33E	501 504 504 504			S09 N32 N29	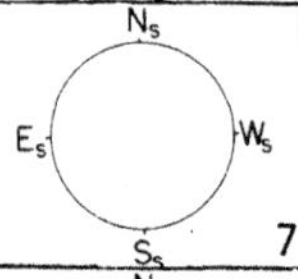	23 7
12 q			7 0.4 1o0o0+0+ 2-2+3+3-	81 160	0 (241) 1	33E - 34E 34^{CE}_{D}	503 504 505 503				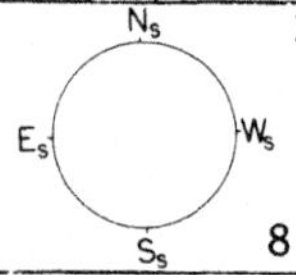	24 8
13 Q			4 0.2 1o1-1+1+ 1+1+1-1-	78 174	0 247 0	33E - 30E 30D	502 503 503 504				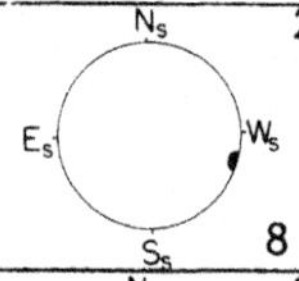	22 8
14			13 0.8 2-2-2o3+ 4-2+3+3o	102 174	0 262 2	33E 31E 30E 32C	501 503 506 504			S09	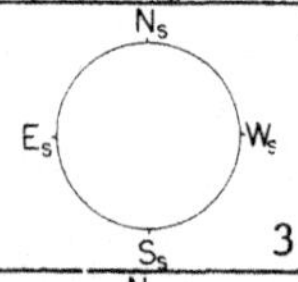	24 3
15			15 1.0 3o3+2+3o 3o3-3+3+	96 176	2 252 1	- 30f 31D 31D	502 503 508 509			S18	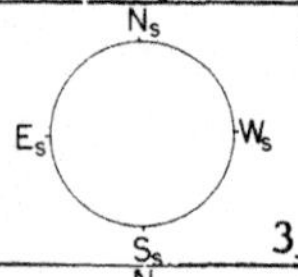	24 3
16 Q			4 0.1 2-1o1-0+ 0+1o2o1-	116 188	0 (223) 0	- - 27C 27D	503 504 510 512			S08	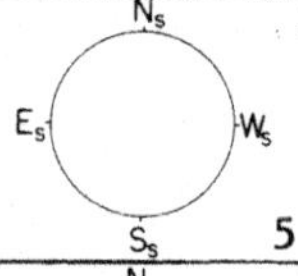	23 5
17			17 1.0 0624 (2) 1o2-3-4+ 4-3o4-4-	107 199	1 (256) 2	- 32E 32E 31D	501 503 507 508			N06 N09	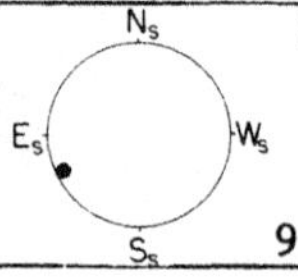	24 9.5

Oct. 8 Slight magnetic activity. Solar electron corona, Climax.

Oct. 9 Slight magnetic activity - one of ten quiet days. U.S.A. rocket AA6.209C; atmospheric composition and upper atmos. physics, 143 km.; 0840 UT; Eglin Field. U.S.A. rocket AA6.213C; atmospheric composition and upper atmos. physics; 116 km.; 1117 UT; Eglin Field.

Oct.10 One of five quiet magnetic days.

Oct.11 One of five quiet magnetic days. Solar electron corona, Climax.

Oct.12 Slight magnetic activity - one of ten quiet days. Bright surge limb, 1400-1500 UT, 15S 90W. U.S.A. rocket AA6.208C; atmospheric composition and upper atmos. physics; 123 km., 0900 UT; Eglin Field. U.S.A. rocket AA6.206C; atmospheric composition and upper atmos. physics; 123 km.; 1112 UT; Eglin Field.

Oct.13 One of five quiet magnetic days. Bright surges at limb, 0706-0802 and 0812-0851 UT, 15S 90W. Solar electron corona, Climax. U.S.A. rocket AA6.214C; atmospheric composition and upper atmos. physics; 116 km.; 1117 UT; Eglin Field. U.S.A. launched 1959 iota (Explorer VII); temperatures, heavy primary cosmic rays, solar UV and x-radiation, corpuscular radiation, earth's radiation balance, micrometeorites, ionosphere; estimated life 20 years.

Oct.14 Slight-to-moderate magnetic activity.

Oct.15 Slight magnetic activity. Solar electron corona, Climax. U.S.A. rocket AA6.207C; atmospheric composition and upper atmos. physics; 122 km.; 1105 UT; Eglin Field.

Oct.16 One of five quiet magnetic days. U.S.A. rocket AA6.215C; atmospheric composition and upper atmos. physics; 139 km.; 1117 UT, Eglin Field.

Oct.17 Slight-to-moderate magnetic activity with sudden commencement at 0624 UT.

Hα flares

Oct. 8 - 2, 1425-1506, N03 W06, 9

Oct.10 - 2, 0329-0624, S17 W38, 4

Oct.13 - 2, 0710E-0805, S16 W86, 3; 2, 0815-0925, S18 W85, 8

Oct.17 - 2, 2130-2332, S21 E64, 1

1959

OCTOBER	GEO-ALERT	SWI	MAGNETIC INDICES Ap C STORM BEGINS / ← 3 HR K_p →	R_z / 3000S	IONO-SPHERE INDEX	AUR-ORAL INDEX	COS-MIC RAY INDEX	TYPE IV OUT-BURST	SWF IMP ≥2 / <2	ACTIVE REGION LAT. / SUNSPOT LAT. AREA	FLARES POSITION IMP ≥2	HRS. PATROL / I_{AREA}
WMI 18	MAG STORM 18/ 01XX		36 1.2 5-4+5-5+ 4+5-4-4+	116 171	2 240 2	34H 30E 31E 31^{D}_{E}	501 504 509 509				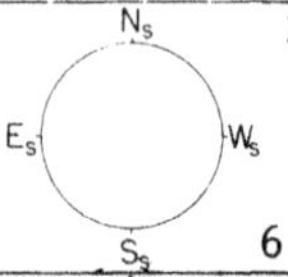	23 6
WMI 19			15 0.9 4-4o2+3+ 2-1+3o3+	111 172	2 - 2	32H 28E 24D 27D	503 505 511 514		1		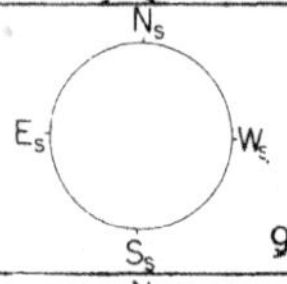	20 9
WMI 20			8 0.4 4-2o3o2o 1+0+0+0+	108 210	1 - 1	26C 23G 29^{D}_{E} 27^{C}_{D}	504 506 513 513		1	S04 S12	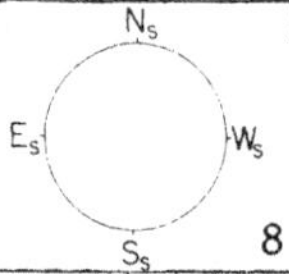	23 8
WMI 21 q			6 0.4 0+1-1-1o 1+3-1+3-	111 205	1 223 1	- - 32D 32D	504 505 513 514			N07 S01	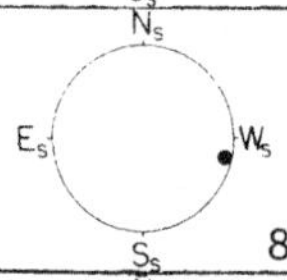	23 8.5
WMI 22			19 1.2 3+4-3+3+ 3+4+3-3-	129 219	2 (222) 3	30H 34G 34^{D}_{E} 34E	504 505 515 510			S10 S13 S16 N09 N03	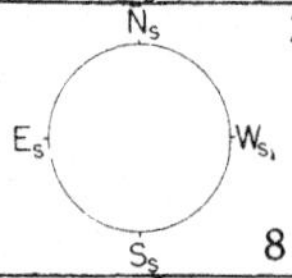	24 8
WMI 23			10 0.5 4-3-3-3+ 2-0+0+0o	135 222	1 211 3	32D 34E 36D 31C	503 506 515 515				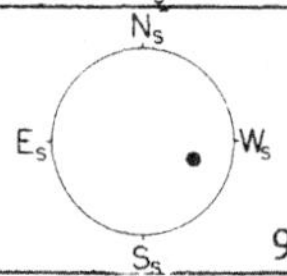	24 9.5
WMI 24 q			6 0.3 0o1-2o2- 2+2o2-2o	143 226	0 222 2	33D 34E 34E 33D	503 506 513 513				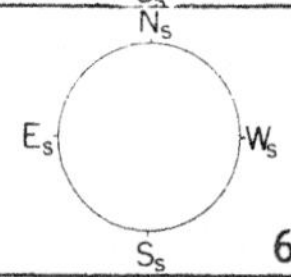	24 6
WMI 25			22 1.2 3o5-4-4+ 4-3o2+4-	137 187	1 - 3	31^{CH}_{G} 34E 36E 33^{D}_{E}	501 504 512 510				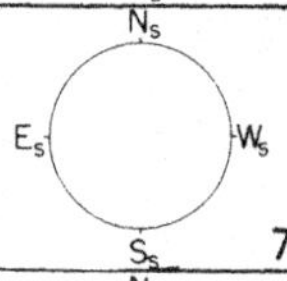	23 7
WMI 26			26 1.2 4-4-4o4o 5o4+3o3+	126 218	2 (228) 3	31D 35G 37D 36D	501 502 510 511			N07 *N07* *950*	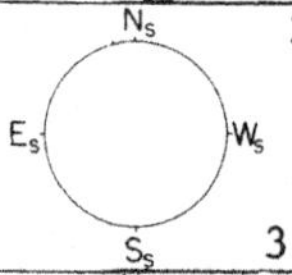	23 3
WMI 27			11 0.7 2+3+4-3o 2+2+1o1o	128 205	2 231 2	34E 34E 34E 34^{D}_{E}	501 503 510 509				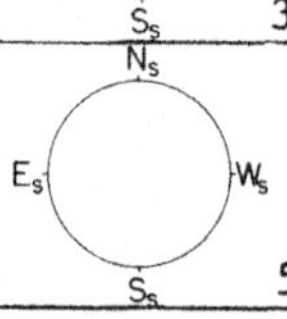	20 5

Oct.18 Moderate magnetic disturbance. Bright surge limb, 1248-1312 UT, 06S 90E. Solar electron corona, Climax.

Oct.19 Slight-to-moderate magnetic activity. Solar electron corona, Climax.

Oct.20 Slight-to-moderate magnetic activity. Active prominence region, 2256-1047 UT (21 Oct.), 42S 90W. Solar electron corona, Climax. U.S.A. rocket AA2.240C; ionospheric physics; 113 km.; 1425 UT; White Sands.

Oct.21 Slight magnetic activity - one of ten quiet days. Bright surge on disk and at limb, 2030-2140 UT, 10S 75W. Sudden disappearance of filaments, 0900-1024 UT, 44S 90W; 1525-1914 UT, 05S 90W.

Oct.22 Slight-to-moderate magnetic activity.

Oct.23 Slight-to-moderate magnetic activity. Outstanding aurora.

Oct.24 Very slight magnetic activity - one of ten quiet days. Solar electron corona, Climax.

Oct.25 Moderate magnetic activity. Outstanding aurora. Bright surge limb, 1255-1320 UT, 10N 90W. U.S.A. rocket AA10.184C; ionospheric physics; 381 km.; 1744 UT; Ft. Churchill.

Oct.26 Moderate magnetic activity. Outstanding aurora. Sudden disappearance of filament, 0807-0924 UT, 18S 90E. Solar electron corona, Climax.

Oct.27 Slight-to-moderate magnetic activity.

Hα flares

Oct.21 - 2, 2057-2140, S10 W69, 3

Oct.23 - 2, 1517-1532D, S08 W34, 2

1959

OCTOBER	GEO-ALERT	SWI	MAGNETIC INDICES Ap C STORM BEGINS ← 3 HR K_p →	R_z 3000 S	IONO-SPHERE INDEX	AUR-ORAL INDEX	COS-MIC RAY INDEX	TYPE IV OUT-BURST	SWF IMP ≥ 2 < 2	ACTIVE REGION LAT. SUNSPOT LAT. AREA	FLARES POSITION IMP ≥2	HRS. PATROL I_{AREA}
28 Q			4 0.0 1-1-2+2- 1o0+0+0o	129 195	0 233 1	33C 30E 30^{CE}_{D} 32C	502 503 510 508		1 1	N06 S09 N10 S10	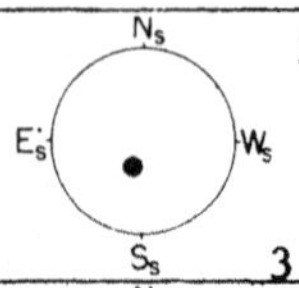	20 3.5
29 q			6 0.2 2347 (57) 0+1+1+1o 2-1-0+3+	129 179	1 246 1	28^{C}_{H} 35D 35D 35D	501 503 509 510			N15	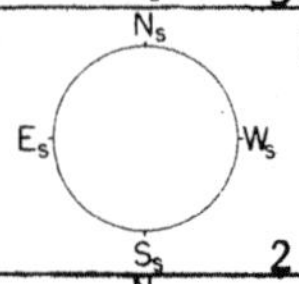	12 2
30	MAG STORM 29/ 2347		28 1.3 5+4o3-2+ 3+3-4o6-	131 174	2 228 2	38D 28E 39^{D}_{E} 39^{D}_{E}	500 500 502 501			N32	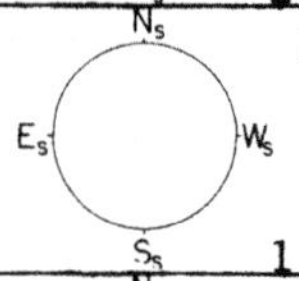	19 1
31 D			38 1.5 4+3o3+5o 4+5o5o6-	141 161	4 213 4	35C 39E 39^{C}_{E} 39^{C}_{E}	498 499 501 499				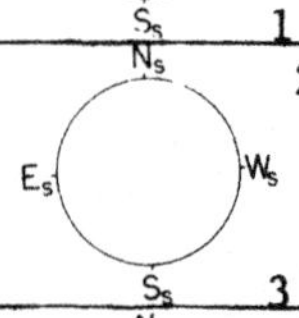	21 3
NOV 1 D			43 1.3 4-5o6+5+ 5+4-4-4-	136 160	6 202 3	35G 39^{D}_{E} 40E 38D	499 502 509 505			N22	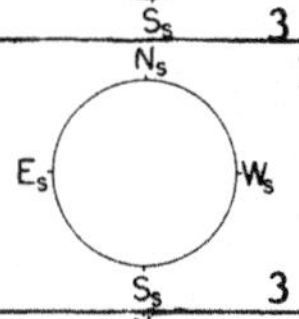	16 3
2 D			69 1.6 5o6-6+5o 6-6-6+6-	121 176	6 199 4	34H 38E 41D 39D	500 502 511 506			N23	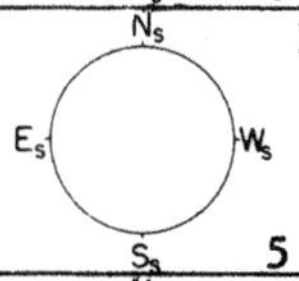	11 5
3 D			48 1.4 6-5-5-5o 5-6-5o5-	97 165	6 - 3	36D 36E 38^{D}_{E} 38D	498 502 508 498			N15	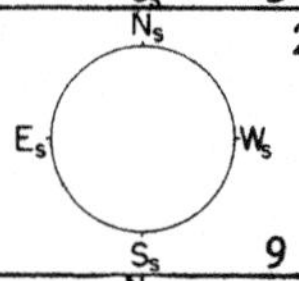	21 9
4			36 1.3 2o3-5-4+ 5o5-6o4+	103 158	5 213 3	33D 38^{D}_{E} 39E 38^{D}_{E}	500 503 506 502		1		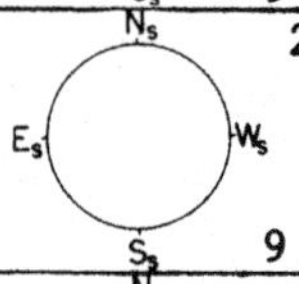	24 9
5			28 1.1 5-5o5o4o 4o3-3o2-	91 166	4 207 3	32D 36E 36E 37D	499 504 508 505		1	N24	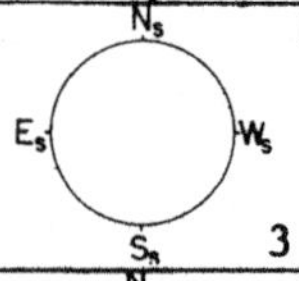	19 3
6			23 1.1 4+4o4-4+ 3+2+3o4+	98 161	4 215 2	31D 34E 34^{D}_{E} 33E	501 503 506 505		3		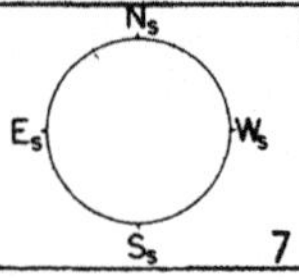	19 7

Oct.28 One of five quiet magnetic days - one of quietest days in IGC.

Oct.29 Slight magnetic activity - one of ten quiet days. Outstanding aurora.

Oct.30 Moderate-to-severe magnetic disturbance. Outstanding aurora.

Oct.31 Moderate-to-severe magnetic disturbance - one of five disturbed days. Outstanding aurora. One of the five disturbed World Blackout days. Solar electron corona, Climax.

Nov. 1 Moderate-to-severe magnetic disturbance - one of five disturbed days. Very outstanding aurora. One of the five disturbed World Blackout days. Solar electron corona, Climax.

Nov. 2 Moderate-to-severe magnetic storm - one of five disturbed days. Very outstanding aurora. One of the five disturbed World Blackout days. Solar electron corona, Climax.

Nov. 3 Moderate-to-severe magnetic storm - one of five disturbed days. Outstanding aurora. One of the five disturbed World Blackout days.

Nov. 4 Moderate-to-severe magnetic disturbance, Outstanding aurora. One of the five disturbed World Blackout days.

Nov. 5 Moderate magnetic activity. Outstanding aurora.

Nov. 6 Slight-to-moderate magnetic activity. U.S.A. rocket AA10.187C; ionospheric physics, 393 km.; 1636 UT, Ft. Churchill. U.S.A. rocket NN2.78C; atmospheric composition; 142 km.; 1925 UT, White Sands.

Hα flares

Oct.28 - 2, 0315E-0340D, S11 E07, 1

1959

NOVEMBER	GEO-ALERT	SWI	MAGNETIC INDICES: Ap, C, STORM BEGINS; ← 3 HR K_p →	R_Z / 3000 S	IONO-SPHERE INDEX	AUR-ORAL INDEX	COS-MIC RAY INDEX	TYPE IV OUT-BURST	SWF IMP ≥2 / <2	ACTIVE REGION LAT. / SUNSPOT LAT AREA	FLARES POSITION IMP ≥2	HRS. PATROL / I_{AREA}
7 q			11 0.7 1+4-3-1o 1+2+4-2+	114 171	3 218 2	27G 31E 32E 29D	503 504 507 508				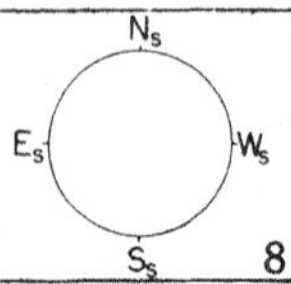	21 8
8			13 0.6 1+4-3-3+ 3+2+3-1+	131 173	2 239 2	30H 29^{fH}_{G} 31E 34E	504 504 506 507				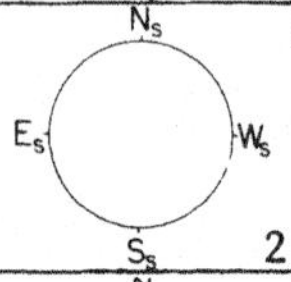	23 2
9 q			9 0.7 1-2o0o1+ 2o3-4-3o	136 187	1 248 1	27C 28E 31E 35^{D}_{E}	500 501 503 504		1		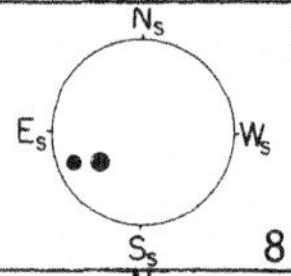	24 8
10 q			10 0.6 3o2+2o2+ 3-1+2o3+	142 197	1 214 1	22A 16G 30E 30E	500 502 503 503		2 2		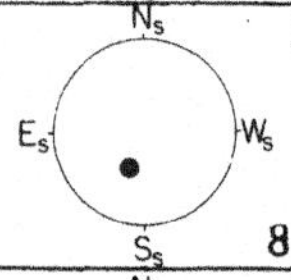	24 8.5
11 Q			5 0.1 2+2-1-2- 1o1-1+2-	153 195	1 235 1	34H 30E 34E 27D	502 502 506 506		1	S16 *S16* *1300*	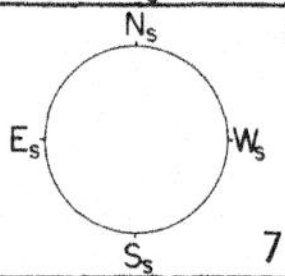	22 7
12 Q			7 0.2 2o2+1-1+ 1o2-3-2+	154 200	0 (235) 0	27D 28E 29E 29E	501 504 509 510				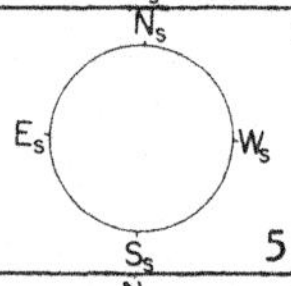	20 5
13			12 0.8 1119 (8) 2+2-2o3- 3+3-3-4-	149 196	1 (228) 1	31D 29^{f}_{H} 31D 32C	502 505 510 510			S15	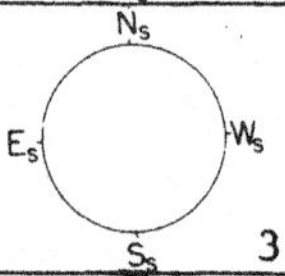	16 3
14			28 1.2 4+4-4o5- 5-4o4-3o	137 195	3 228 2	31D 31^{E}_{H} 33D 31^{C}_{D}	501 505 512 507		3	N19	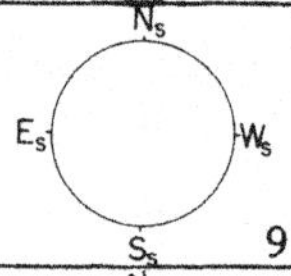	18 9
15 Q			5 0.1 3-2-2-1o 1-1-0o1o	127 184	1 228 1	27D - 23D 27D	503 506 513 509		1		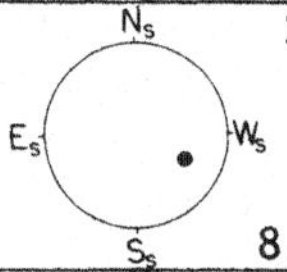	24 8.5
16 q			8 0.5 1-0+2-1+ 2+3+3-3-	113 184	1 267 1	<27D 30E 30E 30E	504 506 513 512			S08 S00 N05	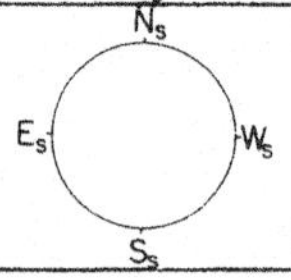	23 7

Nov. 7 Slight-to-moderate magnetic activity but one of ten quiet days.

Nov. 8 Slight-to-moderate magnetic activity.

Nov. 9 Slight-to-moderate magnetic activity but one of ten quiet days. Outstanding aurora. Solar electron corona, Climax.

Nov.10 Slight magnetic activity - one of ten quiet days. Solar electron corona, Climax. U.S.A. rocket OB11.01; ionospheric physics, 1770 km.; 1204 UT; Wallops Island.

Nov.11 One of five quiet magnetic days. Spectral radio emission II, 0657-0714 UT.

Nov.12 Slight magnetic activity - one of five quiet days. Solar electron corona, Climax.

Nov.13 Slight-to-moderate magnetic activity.

Nov.14 Moderate magnetic activity.

Nov.15 Slight magnetic activity - one of five quiet days. Solar electron corona, Climax.

Nov.16 Slight magnetic activity - one of ten quiet days.

Hα flares

Nov. 9 - 2, 0053-0129, S15 E29, 1; 2, 1158E-1212D, S16 E51, 1

Nov.10 - 2, 1104E-1305D, S19 E11, 2

Nov.15 - 2, 0733-0817, 3

1959

NOVEMBER	GEO-ALERT	SWI	MAGNETIC INDICES Ap / C / STORM BEGINS / ← 3 HR K_p →	R_Z / 3000S	IONO-SPHERE INDEX	AUR-ORAL INDEX	COS-MIC RAY INDEX	TYPE IV OUT-BURST	SWF IMP ≥2 / <2	ACTIVE REGION LAT. / SUNSPOT LAT. AREA	FLARES POSITION IMP ≥2	HRS. PATROL / I_{AREA}
(17)			12 0.5 2+4+2+3+ 3+1-0+1-	83 169	2 240 1	28C 32E 32E 28E	503 505 512 509				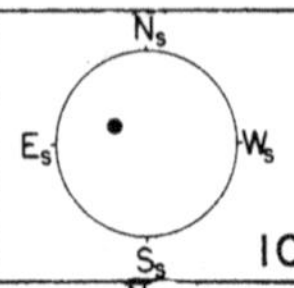	24 10.5
(18)			14 0.9 2-2+2+2+ 3o3o5-2-	73 165	1 198 1	- 29G 30E 33C	504 504 511 512		1		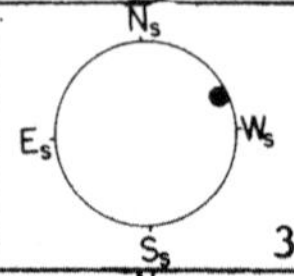	23 3.5
(19)			11 0.5 2o3-4-3+ 3o2+1-1+	69 162	0 230 2	<31C 31$^{E}_{H}$ 30E 31C	504 505 511 510			N33 N13	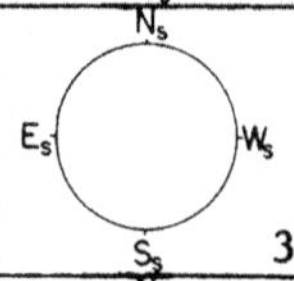	23 3
20 Q			4 0.1 3o2-1+1+ 0o0o0+1-	65 156	1 208 0	<26D 19D 28E <30C	504 505 514 512				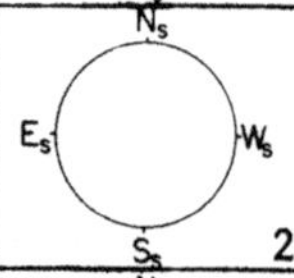	24 2
21			16 1.0 1136 (17) 1o2+2o3+ 4+4o3o3o	70 156	2 205 1	- 28H 38E 35$^{CE}_{D}$	504 506 514 511		1	N07	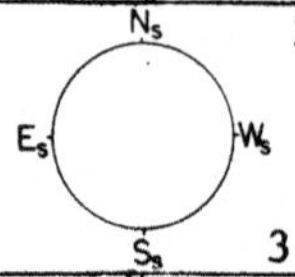	23 3
22			17 1.0 3+4o3+4- 3+2+3+2+	110 174	2 216 2	32C 32E 35E 34E	501 501 508 504				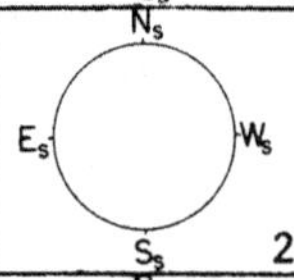	23 2
23	MAG STORM 23/ 0020	S	40 1.4 5+6-5+5o 4+3+4+2o	131 186	4 206 3	35H 38E 38E 38E	502 505 509 502			N05	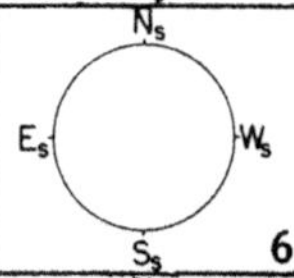	23 6
24 Q		F	8 0.2 3-2o2-1+ 3-2o2-1+	151 202	3 260 2	35D 28G 32C 32C	501 503 506 507			S17	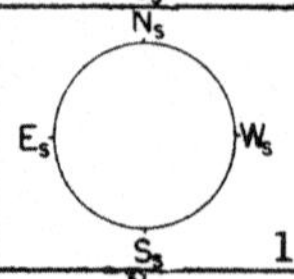	23 1
25 q			10 0.6 2+1+2o3- 3+2o1+3o	162 221	2 253 1	29D 34$^{D}_{E}$ 36E 35E	502 505 508 506			N24 *N15* *1475*	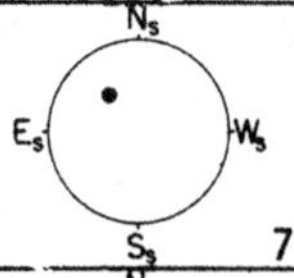	23 7.5
26			13 0.7 2+2o3+4o 3o2-2-3+	161 223	2 265 2	31$^{C}_{D}$ 36E 34E 32$^{D}_{E}$	503 506 510 512		1 1	N16	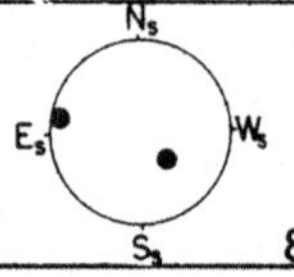	21 8

Nov.17 Slight-to-moderate magnetic activity. Sudden disappearance of filament, 1229-1318 UT, 10S 90W.

Nov.18 Brief moderate magnetic activity. Spectral radio emission II, 0043-0044 UT. U.S.A. rocket AA8.241C; ionospheric physics, 206 km.; 1930 UT, Eglin Field. U.S.A. rocket NASA3.15, upper atmos. physics, 250 km.; 2217 UT; Wallops Island.

Nov.19 Slight-to-moderate magnetic activity. Sudden disappearance of filament 1303-1352 UT, 15N 90W. U.S.A. rocket AA3.282C; geomagnetic field; 106 km.; 0744 UT; Ft. Churchill. U.S.A. rocket AA6.163C; upper atmos. physics; 132 km.; 1036 UT; Ft. Churchill.

Nov.20 One of five quiet magnetic days - one of quietest days in IGC. Solar electron corona, Climax. U.S.A. rocket AA6.164C; upper atmos. physics; 132 km.; 1816 UT; Ft. Churchill.

Nov.21 Slight-to-moderate magnetic activity. Outstanding aurora. Spectral radio emission II, 1741-1752 UT.

Nov.22 Slight-to-moderate magnetic activity. Outstanding aurora.

Nov.23 Moderate-to-severe magnetic disturbance. Outstanding aurora.

Nov.24 Slight magnetic activity - one of five quiet days. Outstanding aurora.

Nov.25 Slight magnetic activity - one of ten quiet days. Outstanding aurora. Sudden disappearance of filament 0935-1130 UT, 30N 25E.

Nov.26 Slight-to-moderate magnetic activity. Outstanding aurora.

Hα flares

Nov.17 - 2+, 0008E-0200D, N14 E22, 1

Nov.18 - 2, 2322-0035D, N23 W66, 2

Nov.25 - 2, 1214-1306D, N26 E22, 1

Nov.26 - 2, 0746-0827, N11 E66, 5; 2+, 0923-1110, S15 W17, 7

1959

NOVEMBER	GEO-ALERT	SWI	MAGNETIC INDICES: Ap / C / STORM BEGINS; 3 HR Kp	R_Z / 3000S	IONO-SPHERE INDEX	AUR-ORAL INDEX	COS-MIC RAY INDEX	TYPE IV OUT-BURST	SWF IMP ≥2 / <2	ACTIVE REGION LAT / SUNSPOT LAT AREA	FLARES POSITION IMP ≥2	HRS. PATROL / I_{AREA}
27			15 0.8 2351 (59) 4-2+3o2o 2o2o3-5-	157 215	3 268 1	31D 32E 36D 36D	502 504 508 510				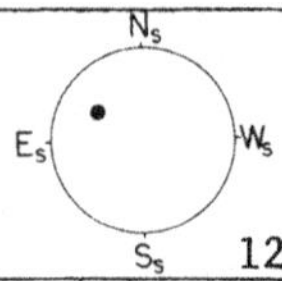	23 12.5
28 D	MAG STORM 27/ 2350	S	82 1.7 6o8o7+7- 4+4-4o4o	151 229	4 215 3	39G 39G 38E 35^{D}_{E}	501 502 505 503		1 1	N23	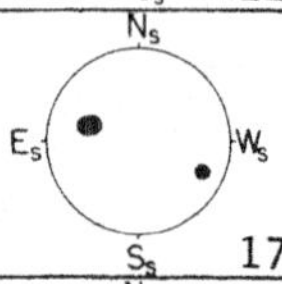	21 17
29		F	15 0.8 4-3-2-4o 3+3o3-2-	161 231	4 242 2	31D 34E 35D 35D	500 502 506 505		2	N22	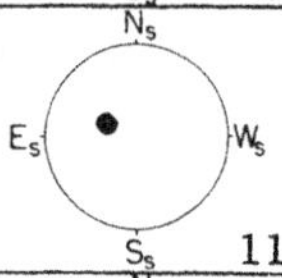	22 11
30 D	MAG STORM 30/ 07XX	S	43 1.4 0553 (6) 0653 (9) 3-2+5+6- 5+6o4o5-	175 230	6 (258) 3	31D 38E 38E 38E	500 503 508 508	Syd Har	2		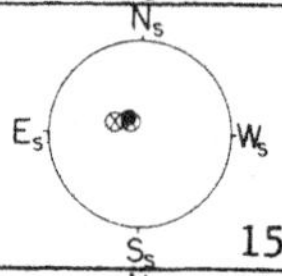	23 15
DEC 1		C	28 1.1 5o4+4+5- 4+3+3o2-	170 222	5 (253) 2	34D 33^{D}_{E} 35E 38D	504 508 510 510	Har	3	N09 *N08* *2275*	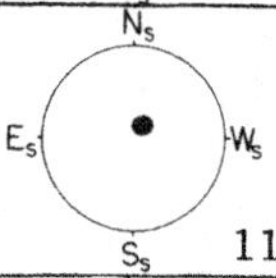	22 11.5
2		F	30 1.2 3+4+4+5o 4o4+4o4o	165 217	7 232 2	34D 33D 36D 38D	500 507 511 509		3	N22	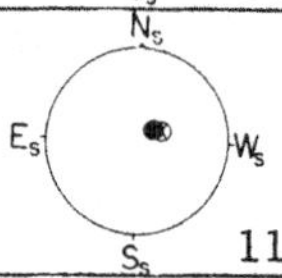	16 11
3 D			50 1.5 5o5-5o5+ 5+6-4o5+	160 202	6 (215) 3	33D 34^{CH}_{E} 38E 37^{D}_{E}	495 495 489 489		2 1		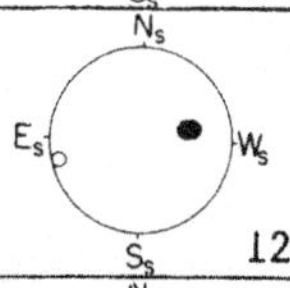	22 12.5
4			12 0.7 3+4o3-3- 2o2-2+2+	163 199	4 (275) 2	35D 37D 37D 37D	495 494 479 483		2 2		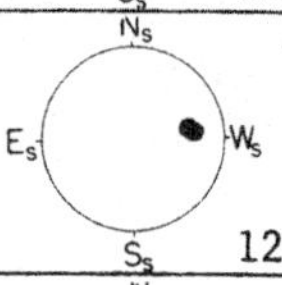	22 12.5
5 D	MAG STORM 05/ 0659		68 1.7 0659 (57) 0o0o4o5- 6+7-8-6+	142 204	4 218 3	33^{C}_{D} 34G 46D 46D	499 498 488 488		3 1	N12 *N10* *1100*	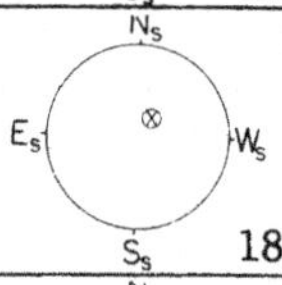	22 18.5
6			16 0.9 3o4+3+3+ 3o2o3o2+	147 202	2 (226) 2	38C 33E 36D 35E	499 499 492 487				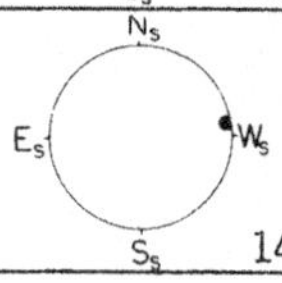	24 14.5

Nov.27 Moderate magnetic disturbance begins with sudden commencement at 2351 UT. Outstanding aurora.

Nov.28 Severe magnetic storm continues - one of five disturbed days. Outstanding aurora. Spectral radio emission II, 2017-2045 UT.

Nov.29 Slight-to-moderate magnetic activity. Outstanding aurora. Spectral radio emission II, 1854-1904 UT.

Nov.30 Moderate-to-severe magnetic storm with sudden commencements at 0553 and 0653 UT - one of five disturbed days. Outstanding aurora. One of the five disturbed World Blackout days. Active dark filament, 1744-1815 UT, 17S 13W. Spectral radio emission II 0251-0358 UT; IV 0312-0350 UT; IV 1739-1940 UT; II 1741-1810 UT; IV 2030-2045 UT; IV 2142-2330 UT. U.K. rocket Skylark - 16; sodium vapor trail and atmospheric pressure; 1046 UT; Woomera. U.K. rocket Skylark - 60; grenades (temp., winds); 1305 UT; Woomera.

Dec. 1 Moderate magnetic disturbance. Outstanding aurora. Spectral radio emission IV, 1400-1517, 1827-1837 and 2204-2240 UT; II, 1733-1734 UT. U.K. rocket Skylark - 38; grenades (temp; winds) and dielectric (electron densities); 1206 UT; Woomera.

Dec. 2 Moderate magnetic disturbance. Outstanding aurora. One of the five disturbed World Blackout days.

Dec. 3 Moderate-to-severe magnetic disturbance - one of five disturbed days. Outstanding aurora. One of the five disturbed World Blackout days.

Dec. 4 Slight-to-moderate magnetic activity. Outstanding aurora. Dark surges on disk, 1848-1934, 2016-2034, 2036-2116 and 2235-2256 UT, 08N 02W

Dec. 5 Severe magnetic storm begins with sudden commencement at 0659 UT one of five disturbed days. Very outstanding aurora. Dark surges on disk frequent all day, 10N 10W.

Dec. 6 Slight-to-moderate magnetic activity. Outstanding aurora.

Hα flares

Nov.27 - 2, 1220-1337D, N19 E32, 3
Nov.28 - 2, 1207-1335, N11 E34, 4; 2, 1335E-1340D, S17 W46, 1
3, 2006-2130, N12 E31, 3
Nov.29 - 2, 1335E-1502, N10 E20, 3; 2+, 1816-2012, N09 E18, 4
Nov.30 - 2+, 0247-0356, N08 E16, 1; 2, 1111-1150, N11 E07, 2;
3, 1720-1906, N07 E06, 5
Dec. 1 - 2, 1056E-1112D, N08 W04, 1; 2, 1456E-1622, N09 W06, 3;
2+, 1638-2035, N09 W05, 4
Dec. 2 - 2, 0500E-0656, N08 W10, 2; 2+, 1219E-1412D, N07 W16, 2
Dec. 3 - 2, 0715-0743, S13 E67, 1; 2, 1010-1107, N09 W30, 2;
2, 1756-2000, N09 W34, 2
Dec. 4 - 2, 0341-0412, N07 W40, 2; 2, 0732-1005, N07 W38, 4;
2+, 1814-2100, N06 W44, 2
Dec. 5 - 2, 1215-1250, N11 W07, 2
Dec. 6 - 2, 0748-0810, N07 W76, 1

1959

DECEMBER	GEO-ALERT	SWI	MAGNETIC INDICES Ap C STORM BEGINS ← 3 HR K_p →	R_z 3000 S	IONO-SPHERE INDEX	AUR-ORAL INDEX	COS-MIC RAY INDEX	TYPE IV OUT-BURST	SWF IMP ≥ 2 < 2	ACTIVE REGION LAT. SUNSPOT LAT. AREA	FLARES POSITION IMP ≥2	HRS. PATROL I_{AREA}
7 Q			6 0.3 3o3-1-0+ 2-2-1-1+	141 191	1 186 1	32D >25E 33E 30^C_E	497 499 497 495		2	S15	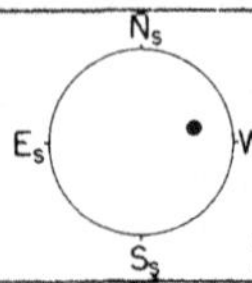	21 10.5
8 q			7 0.3 1+2o2-1o 1-3-2+2o	145 193	1 - 0	31^D_H 32E 32E 33E	497 498 495 493		2	N18 S15	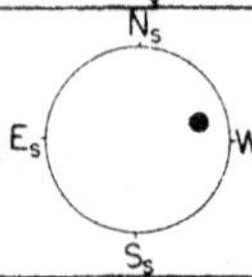	14 7.5
9 q			8 0.3 1+1+3-3- 3o2o1-2o	94 173	2 - 1	32D 29H 33E 34E	499 501 502 502			N08	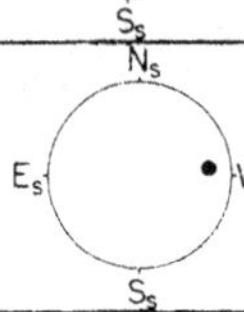	20 7.5
10 Q			6 0.2 2+2-2-1- 0+1+2+1+	89 174	2 242 1	28D 15G 30E 31D	498 502 505 504		1 1		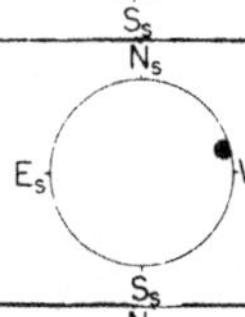	21 9
11 Q			6 0.2 0+2o2-2o 1+2+1-2-	82 171	2 (237) 0	30D 25^G_H 30^{CE}_D 31^C_E	499 501 505 504		1	N05 N29	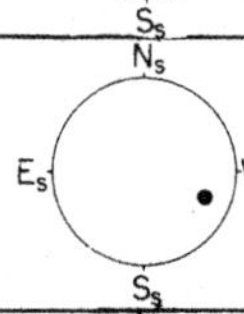	22 7.5
RWI 12			16 1.0 2+4+3-3o 3-2o3o4o	75 169	1 228 1	27B 29E 30E 32D	499 502 506 503		1		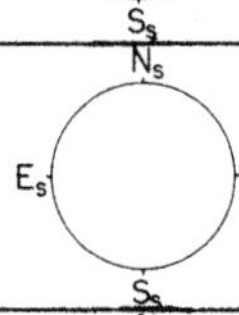	22 3
RWI 13			18 1.1 4o3+2+2- 1o2+3o5+	73 162	2 219 2	31D 29G 30^D_E 34D	500 505 509 506				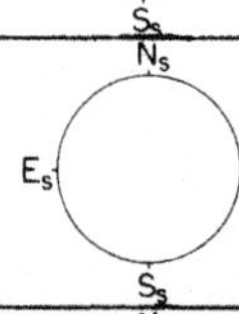	20 3
RWI (14) D			40 1.4 5-4+5+5o 5-5+5-3o	88 165	5 230 3	31H 33H 37E 35E	502 505 510 503			S09	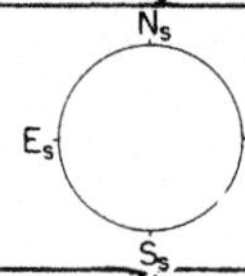	22 3
RWI (15)			26 1.1 3+4+4+4o 3+3+4o4+	123 171	5 - 2	29G 30^D_E 34E 34D	503 505 509 503			N18 S14 *N19* *1625*	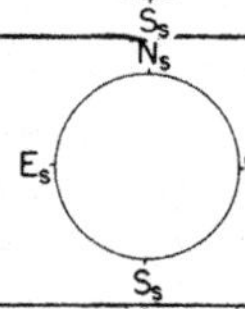	23 8
RWI (16)			17 1.0 3+2+3o3+ 4-4o2+3o	113 164	4 - 2	34^D_E 30E 34E 33E	502 506 511 508			N07	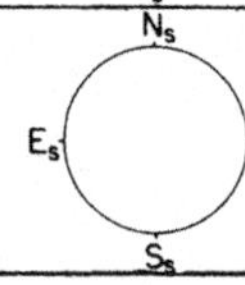	22 3

Dec. 7 Slight magnetic activity - one of five quiet days. Solar electron corona, Climax.

Dec. 8 Slight magnetic activity - one of ten quiet days. Spectral radio emission II, 0121-0126 UT.

Dec. 9 Slight magnetic activity - one of ten quiet days.

Dec.10 One of five quiet magnetic days. Spectral radio emission II, 0521-0529 UT.

Dec.11 One of five quiet magnetic days. Spectral radio emission II, 0412-0418 UT. Solar electron corona, Climax.

Dec.12 Slight-to-moderate magnetic activity. Solar electron corona, Climax.

Dec.13 Moderate magnetic activity. Solar electron corona, Climax.

Dec.14 Moderate magnetic disturbance - one of five disturbed days. Outstanding aurora. Solar electron corona, Climax.

Dec.15 Slight-to-moderate magnetic activity. One of the five disturbed World Blackout days.

Dec.16 Slight-to-moderate magnetic activity. Solar electron corona, Climax.

Hα flares

Dec. 7 - 2, 2134-2220D, N09 W37, 3

Dec. 8 - 2, 0747-0839, N12 W44, 2

Dec. 9 - 2, 0554-0658, N05 W50, 1

Dec.10 - 2, 0102-0150, N15 W68, 1; 2+, 0512-0537D, N15 W70, 1

Dec.11 - 2, 0910E-0957, S16 W44, 2

1959

DECEMBER	GEO-ALERT	SWI	MAGNETIC INDICES Ap C STORM BEGINS ← 3 HR K_p →	R_Z 3000 S	IONO-SPHERE INDEX	AUR-ORAL INDEX	COS-MIC RAY INDEX	TYPE IV OUT-BURST	SWF IMP ≥ 2 < 2	ACTIVE REGION LAT. SUNSPOT LAT. AREA	FLARES POSITION IMP ≥2	HRS. PATROL I_{AREA}
RWI (17) q			8 0.3 3o1+3-3- 2o1-1+2-	107 167	1 232 1	24G 29^{E}_{G} 33D 25E	504 508 514 506		1	S03	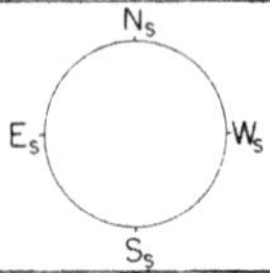	23 4
RWI 18			10 0.5 2o2-3o3- 4-2o1+2o	117 169	2 210 1	22A 31f 32E 34D	503 505 512 511	Syd	1		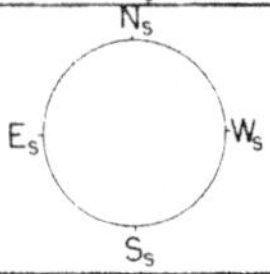	24 2
RWI 19			14 0.8 2+3-3+4- 4o3-2o1o	134 180	3 210 2	30G 30^{EH}_{G} 34D 33E	503 504 509 509		1	N10	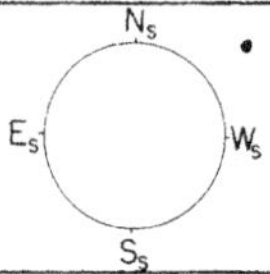	19 3
RWI 20 q			6 0.3 2+2o1+2- 1+2o2-2-	133 178	0 C 1	30H 25E 30^{D}_{E} 34^{D}_{E}	504 506 514 515				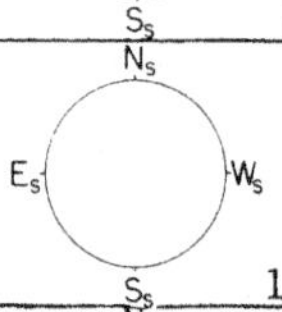	23 11
RWI 21 Q			4 0.1 1+1+1+1- 1-0+1o1o	131 185	1 (237) 1	34E 26H 34C 35E	504 507 517 518	Syd		S17	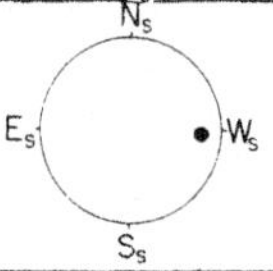	17 8.5
22 Q			6 0.4 2-0+0+0+ 1o2+2o3o	126 171	2 (273) 1	30D 28E 36D 33^{D}_{E}	502 504 514 518			N11	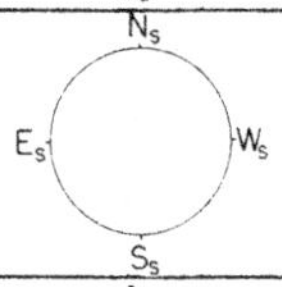	22 1
23			28 1.3 1525 (49) 3-3-4-3- 3-6o4+5-	122 166	2 223 2	32C 30E 35E 36D	500 502 506 506			N03 N24 N17	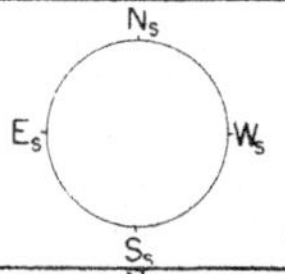	19 3
24			15 0.9 4o1+4o4o 3+1+2-2o	116 163	2 246 2	33C 32E 35^{D}_{E} 35E	500 501 502 499		1	N25	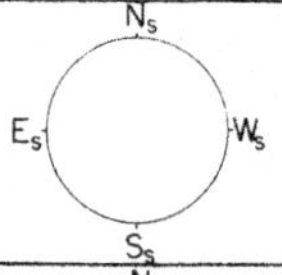	12 1
25			8 0.4 2o3o2o2- 2-2-2+1+	121 -	1 240 2	32C 30E 34E 37E	497 499 498 494			N20	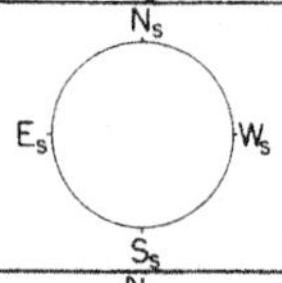	18 4
26			26 1.2 1022 (2) 3-4-3+4+ 4+4-4o5-	124 161	4 - 2	31D 30E 34^{D}_{E} 36E	500 504 508 507			N09 N09 810	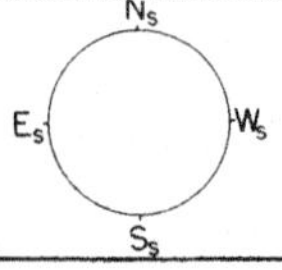	19 1

Dec.17 Slight magnetic activity - one of ten quiet days.

Dec.18 Slight-to-moderate magnetic activity. Spectral radio emission II, 0640-0647 UT; IV, 0653-0723 UT. Solar electron corona, Climax.

Dec.19 Slight-to-moderate magnetic activity.

Dec.20 Very slight magnetic activity - one of ten quiet days. Solar electron corona, Climax.

Dec.21 One of five quiet magnetic days - one of quietest days of IGC. Outstanding aurora. Bright surge limb, 0053-0133 UT, 05S 90W. Spectral radio emission II, 0055-0120 UT; IV, 0120-0350 UT.

Dec.22 Slight magnetic activity but one of five quiet days. Outstanding aurora.

Dec.23 Moderate magnetic storm begins with sudden commencement at 1525 UT. Outstanding aurora.

Dec.24 Slight-to-moderate magnetic activity. Outstanding aurora.

Dec.25 Slight magnetic activity. Outstanding aurora.

Dec.26 Moderate magnetic disturbance with sudden commencement at 1022 UT. Outstanding aurora.

Hα flares

Dec.21 - 2, 0043-0122, S04 W53, 4

1959

DECEMBER	GEO-ALERT	SWI	MAGNETIC INDICES Ap C STORM BEGINS ← 3 HR K_p →	R_Z 3000S	IONO-SPHERE INDEX	AUR-ORAL INDEX	COS-MIC RAY INDEX	TYPE IV OUT-BURST	SWF IMP ≥2 <2	ACTIVE REGION LAT. SUNSPOT LAT. AREA	FLARES POSITION IMP ≥2	HRS. PATROL I_{AREA}
27 D	MAG STORM 26/ 11XX		38 1.4 5-5-4o6- 4+5-4o4+	132 167	6 - 3	31D 33E 38D 38D	501 505 508 510				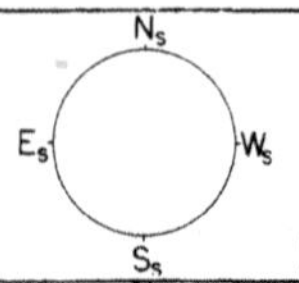	22 5
28 D			40 1.4 5o5-4+5- 5+5+5-3o	127 172	5 221 2	33D 34^{D}_{E} 38D 37D	502 505 505 503				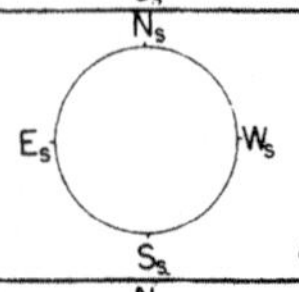	23 4
29			18 0.9 4o4o4o4- 4-3-1o2o	136 171	4 - 2	33D 34E 34E 34E	502 504 504 503			N10	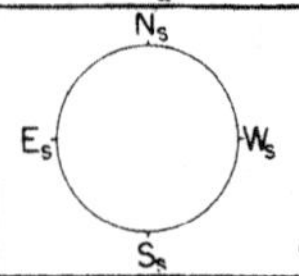	23 4
30			14 0.8 2-3o3o4- 4-3o3-1o	127 179	4 - 2	31D 30E 36E 36E	501 501 0 500			N25 S08	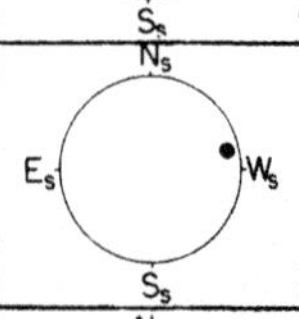	23 3.5
31 q			7 0.4 2-2-3-2o 1o1o2+1+	153 167	1 231 2	15H 28^{E}_{H} 33D 32^{D}_{E}	501 505 512 505			N09	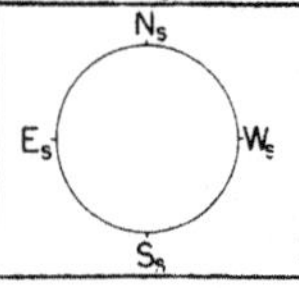	24 3

Dec.27 Moderate-to-severe magnetic storm continues - one of five disturbed days. Outstanding aurora. One of the five disturbed World Blackout days.

Dec.28 Moderate magnetic disturbance - one of five disturbed days. Outstanding aurora. One of the five disturbed World Blackout days.

Dec.29 Slight-to-moderate magnetic activity. Sudden disappearance of filament 1340-1444 UT, 30S 90E. Solar electron corona, Climax.

Dec.30 Slight-to-moderate magnetic activity. Outstanding aurora. Solar electron corona, Climax.

Dec.31 Slight magnetic activity - one of ten quiet days.

Hα flares

Dec.30 - 2, 0730E-0800D, N10 W60, 1

SUPPLEMENTARY TABLES

SUPPLEMENTARY TABLE 1.

International Geophysical Cooperation 1959 Final Calendar

1959 JANUARY 1959

Sun.	Mon.	Tue.	Wed.	Thu	Fri.	Sat.
				1	2	3
4	5	6	7	8	9	10
11	12	13	14	15	16	17
18	19	20	21	22	23	24
25	26	27	28	29	30	31

1959 FEBRUARY 1959

Sun.	Mon.	Tue.	Wed.	Thu	Fri	Sat.
1	2	3	4	5	6	7
8	9	10	11	12	13	14
15	16	17	18	19	20	21
22	23	24	25	26	27	28

1959 MARCH 1959

Sun.	Mon.	Tue.	Wed.	Thu	Fri	Sat.
1	2	3	4	5	6	7
8	9	10	11	12	13	14
15	16	17	18	19	20	21
22	23	24	25	26	27	28
29	30	31				

1959 APRIL 1959

Sun.	Mon.	Tue.	Wed.	Thu	Fri.	Sat.
			1	2	3	4
5	6	7	8	9	10	11
12	13	14	15	16	17	18
19	20	21	22	23	24	25
26	27	28	29	30		

1959 MAY 1959

Sun.	Mon.	Tue.	Wed.	Thu	Fri.	Sat.
					1	2
3	4	5	6	7	8	9
10	11	12	13	14	15	16
17	18	19	20	21	22	23
24/31	25	26	27	28	29	30

1959 JUNE 1959

Sun.	Mon.	Tue.	Wed.	Thu.	Fri	Sat.
	1	2	3	4	5	6
7	8	9	10	11	12	13
14	15	16	17	18	19	20
21	22	23	24	25	26	27
28	29	30				

1959 JULY 1959

Sun.	Mon.	Tue.	Wed.	Thu	Fri.	Sat.
			1	2	3	4
5	6	7	8	9	10	11
12	13	14	15	16	17	18
19	20	21	22	23	24	25
26	27	28	29	30	31	

1959 AUGUST 1959

Sun.	Mon.	Tue.	Wed.	Thu	Fri	Sat.
						1
2	3	4	5	6	7	8
9	10	11	12	13	14	15
16	17	18	19	20	21	22
23/30	24/31	25	26	27	28	29

1959 SEPTEMBER 1959

Sun.	Mon.	Tue.	Wed.	Thu	Fri	Sat.
		1	2	3	4	5
6	7	8	9	10	11	12
13	14	15	16	17	18	19
20	21	22	23	24	25	26
27	28	29	30			

1959 OCTOBER 1959

Sun.	Mon.	Tue.	Wed.	Thu	Fri.	Sat.
				1	2	3
4	5	6	7	8	9	10
11	12	13	14	15	16	17
18	19	20	21	22	23	24
25	26	27	28	29	30	31

1959 NOVEMBER 1959

Sun.	Mon.	Tue.	Wed.	Thu	Fri.	Sat.
1	2	3	4	5	6	7
8	9	10	11	12	13	14
15	16	17	18	19	20	21
22	23	24	25	26	27	28
29	30					

1959 DECEMBER 1959

Sun.	Mon.	Tue.	Wed.	Thu	Fri	Sat.
		1	2	3	4	5
6	7	8	9	10	11	12
13	14	15	16	17	18	19
20	21	22	23	24	25	26
27	28	29	30	31		

1960 JANUARY 1960

Sun.	Mon.	Tue.	Wed.	Thu.	Fri.	Sat.
					1	2
3	4	5	6	7	8	9
10	11	12	13	14	15	16
17	18	19	20	21	22	23
24/31	25	26	27	28	29	30

(3) Regular World Day
(18) Regular World Day with highest priority
(9) Regular World Day with unusual meteoric activity
(10) Regular World Day with highest priority and unusual meteoric activity
[2] Day of total eclipse

17 18 19 20 21 / 22 23 24 25 26 — Regular World Interval (R W I)

16 17 18 / 19 20 21 22 23 24 25 — World Meteorological Interval (W M I)

SUPPLEMENTARY TABLE 2

1959 GEOPHYSICAL ALERTS

PERIODS OF SOLAR ALERT

Number	Started 1600 UT	Ended 1600 UT
41	Jan. 1	Jan. 3
42	Jan. 10	Jan. 14
43	Jan. 22	Jan. 29

GEOPHYSICAL ALERTS AT 1600 UT

Number	Date Issued	Type	Based on event at Date/Time UT
1	Feb. 11	Magstorm	11/0756
2	Feb. 25	Magstorm (Aurora probable)	25/0215
3	Mar. 26	Magstorm (Aurora probable)	26/0840
4	Apr. 10	Magstorm	09/1827
5	May 12	Magstorm	11/2318
6	May 16	Magstorm	15/1100
7	May 24	Magstorm	24/0540
8	June 11	Magstorm	11/0910
9	June 28	Magstorm	27/1100
10	June 29	Magstorm	29/0728
11	July 12	Magstorm	11/1623
12	July 15	Magstorm (Aurora probable)	15/0802
13	July 18	Magstorm (Aurora probable)	17/1638
14	Aug. 16	Magstorm	15/1400
15	Aug. 20	Magstorm	20/0410
16	Sep. 2	Magstorm	02/0110
17	Sep. 4	Magstorm (Aurora probable)	03/2159
18	Sep. 19	Magstorm	18/21XX
19	Sep. 25	Magstorm	25/0045
20	Oct. 1	Magstorm	01/0430
21	Oct. 4	Magstorm (Aurora probable)	03/17XX
22	Oct. 18	Magstorm	18/01XX
23	Oct. 30	Magstorm	29/2347
24	Nov. 23	Magstorm	23/0020
25	Nov. 28	Magstorm (Aurora probable)	27/2350
26	Nov. 30	Magstorm	30/07XX
27	Dec. 5	Magstorm	05/0659
28	Dec. 27	Magstorm	26/11XX

SUPPLEMENTARY TABLE 3

1959 SPECIAL WORLD INTERVALS

Number	Started 0001 UT	Ended 2359 UT
22	January 24	January 25

Number	Started upon receipt after 1600 UT	Ended 2359 UT
1	Feb. 11	Feb. 12
2*	Feb. 25	Feb. 26
3	Mar. 26	Mar. 28
4	Apr. 10	Apr. 11
5	May 12	May 13
6	June 11	June 12
7	July 15	July 16
8	Sep. 4	Sep. 5
9	Sep. 20	Sep. 21
10	Nov. 23	Nov. 24
11	Nov. 28	Nov. 29
12	Nov. 30	Dec. 2

*Sent to NIZMIR (U.S.S.R.) and broadcast on WWV/H but through oversight it was not issued to the World Meteorological Network or other telegraphic addresses.

SUPPLEMENTARY TABLE 4A

WORLD BLACKOUT INDEX (P)

1959

Date	Jan.	Feb.	Mar.	Apr.	May	June	July	Aug.	Sept.	Oct.	Nov.	Dec.
1	1	2	5	2	2	1	1	4	4	3	6	5
2	0	2	5	0	1	0	2	4	3	1	6	7
3	0	1	4	1	0	1	1	4	3	4	6	6
4	1	3	3	1	1	2	2	3	6	5	5	4
5	2	2	4	0	0	1	3	2	5	4	4	4
6	4	2	2	1	0	1	2	3	1	5	4	2
7	2	1	1	0	0	0	2	2	1	4	3	1
8	1	1	1	2	2	1	2	2	1	1	2	1
9	3	2	0	3	1	2	4	3	1	1	1	2
10	4	1	0	5	2	1	7	2	1	1	1	2
11	2	3	1	2	8	2	8	1	1	0	1	2
12	1	3	1	1	9	1	9	0	1	0	0	1
13	1	4	1	0	9	1	8	0	1	0	1	2
14	0	6	0	1	8	2	8	1	2	0	3	5
15	0	7	1	1	7	3	9	1	2	2	1	5
16	3	5	0	1	8	2	9	6	2	0	1	4
17	2	5	0	1	4	1	9	7	1	1	2	1
18	3	0	0	0	3	1	9	5	1	2	1	2
19	2	2	0	0	3	0	8	6	5	2	0	3
20	1	1	0	0	2	0	6	5	6	1	1	0
21	0	1	0	1	1	0	4	5	7	1	2	1
22	0	1	0	0	2	1	2	4	5	2	2	2
23	0	3	0	3	1	0	2	3	4	1	4	2
24	0	0	1	3	3	2	4	2	4	0	3	2
25	1	4	2	4	1	0	5	2	5	1	2	1
26	2	5	4	3	1	2	4	1	5	2	2	4
27	2	3	7	3	1	3	4	1	4	2	3	6
28	0	5	6	2	1	5	2	0	2	0	4	5
29	1	-	6	3	1	6	2	1	2	1	4	4
30	1	-	5	3	1	4	2	1	1	2	6	4
31	2	-	4	-	1	-	3	1	-	4	-	1
five	6	14	2	10	11	15	12	16	4	3	1	2
most	9	15	27	25	12	27	15	17	20	4	2	3
disturbed	10	16	28	27	13	28	16	18	21	5	3	15
days	16	26	29	29	14	29	17	19	22	6	4	27
	18	28	30	30	16	30	18	21	25	31	30	28

SUPPLEMENTARY TABLE 4B

BLACKOUT INDICES FOR THE NORTH AND SOUTH HEMISPHERES

1959

Date	Jan.		Feb.		Mar.		Apr.		May		June		July		Aug.		Sep.		Oct.		Nov.		Dec.	
	N	S	N	S	N	S	N	S	N	S	N	S	N	S	N	S	N	S	N	S	N	S	N	S
1	2	0	2	1	5	5	3	0	2	2	1	1	1	1	5	2	5	3	4	1	6	5	6	2
2	0	0	2	2	6	4	0	0	1	0	0	0	3	0	5	2	4	2	2	0	7	5	7	6
3	0	0	1	2	5	3	1	1	0	0	1	0	1	0	5	1	4	2	5	3	7	4	6	5
4	1	1	3	3	4	2	1	0	1	0	2	1	3	1	4	2	6	6	6	2	6	4	4	3
5	2	2	2	3	4	3	0	0	0	0	1	0	3	2	3	0	6	2	5	3	5	2	4	4
6	4	4	3	1	2	1	1	0	0	0	1	1	2	2	4	1	1	1	6	2	5	2	3	0
7	3	1	1	1	1	2	0	0	0	0	0	0	2	3	3	0	0	1	4	3	4	1	1	1
8	2	0	1	1	1	1	3	0	2	1	0	1	2	3	2	1	1	0	2	0	3	0	0	2
9	3	3	2	1	0	0	3	3	0	2	3	0	5	3	4	1	1	1	1	0	1	1	1	4
10	4	4	0	1	0	0	6	3	2	0	1	1	8	5	2	1	0	1	2	0	1	0	1	4
11	2	1	2	4	0	1	3	0	8	8	2	3	9	9	1	0	1	1	0	0	0	1	1	3
12	1	0	3	2	1	1	1	0	9	9	0	2	9	9	0	0	1	1	0	0	0	0	1	1
13	1	0	4	3	1	1	0	0	8	9	0	3	8	9	0	0	1	1	0	0	0	2	2	1
14	0	0	6	7	0	0	1	0	8	8	1	4	8	8	1	0	2	1	0	0	3	2	6	4
15	0	0	6	8	1	0	1	1	8	6	2	5	9	9	1	2	2	1	3	0	2	0	6	3
16	3	3	5	4	0	0	1	0	8	7	1	3	9	9	6	7	3	1	0	0	1	0	5	2
17	3	0	4	7	0	0	1	0	4	4	0	2	9	9	7	7	1	1	0	1	3	1	2	0
18	4	1	0	0	0	0	0	0	5	0	0	1	9	9	6	4	1	1	2	1	2	0	3	0
19	3	1	2	2	0	0	0	0	4	1	0	0	8	8	5	7	4	6	2	0	3	1	4	2
20	1	0	0	4	0	0	0	0	3	0	0	0	8	3	3	7	6	7	1	0	1	1	0	0
21	0	0	0	1	0	0	1	0	2	0	0	0	5	2	6	4	7	6	0	1	2	1	1	0
22	0	0	1	1	0	0	0	0	2	1	1	0	3	0	5	2	6	4	2	1	3	1	2	1
23	0	0	3	2	0	0	2	4	1	1	0	0	3	1	4	2	5	1	1	1	5	2	2	1
24	0	0	0	0	1	0	1	7	3	4	2	1	4	3	3	0	4	3	0	0	4	2	2	1
25	1	1	3	5	2	2	3	6	1	1	0	1	6	3	2	1	6	3	1	1	2	2	1	1
26	1	4	5	4	3	5	3	2	1	0	3	1	5	2	1	0	6	2	3	0	3	1	4	3
27	1	3	2	4	7	8	4	1	1	0	4	2	6	1	1	0	5	2	3	1	3	2	6	5
28	0	0	5	5	6	7	2	2	0	1	6	4	3	1	0	0	3	0	0	0	5	3	6	3
29	1	0	-	-	7	5	4	2	0	2	6	5	2	1	1	2	3	0	1	0	5	2	4	3
30	0	1	-	-	6	3	4	1	0	1	4	4	1	3	1	2	2	0	1	3	6	7	4	3
31	2	2	-	-	4	4	-	-	1	1	-	-	3	2	1	1	-	-	4	3	-	-	1	0

Note: The reliability of the index for the South hemisphere is limited by lack of data.

SUPPLEMENTARY TABLE 5

DAILY VALUES OF E-LAYER INDEX J_E

1959

Date	Jan	Feb	Mar	Apr	May	June	July	Aug	Sept	Oct	Nov	Dec
1	(312)	327	235	269	(288)	(294)	(274)	(263)	(344)	237	202	(253)
2	335	343	230	280	256	268	(261)	270	346	259	199	232
3	308	316	244	308	(268)	(262)	(277)	303	(328)	233	-	(215)
4	349	351	265	(275)	282	-	-	(301)	307	187	213	(275)
5	(325)	285	250	288	(286)	-	-	(305)	287	204	207	218
6	327	289	278	310	294	-	(256)	(284)	286	208	215	(226)
7	270	272	237	(313)	-	(267)	292	-	291	224	218	186
8	298	C	263	(304)	(329)	264	234	(275)	(287)	220	239	-
9	288	303	302	275	319	(303)	(305)	(312)	276	231	248	-
10	260	288	299	268	C	312	(288)	-	258	234	214	242
11	287	295	(312)	278	(333)	357	C	(281)	275	244	235	(237)
12	288	264	313	261	298	(324)	C	303	272	(241)	(235)	228
13	313	261	293	304	(321)	-	(338)	275	(270)	247	(228)	219
14	(276)	233	C	294	(305)	(308)	-	-	284	262	228	230
15	315	243	314	271	(295)	-	-	(289)	289	252	228	-
16	316	251	(318)	293	C	(317)	-	C	264	(223)	267	-
17	(285)	225	(282)	(304)	C	-	-	257	269	(256)	240	232
18	-	274	319	269	(349)	302	-	-	280	240	198	210
19	(287)	277	312	320	-	-	-	(268)	285	-	230	210
20	291	258	(313)	321	(303)	(284)	-	(289)	294	-	208	C
21	374	274	352	301	303	(284)	-	299	260	223	205	(237)
22	(319)	294	372	312	268	-	-	-	253	(222)	216	(273)
23	(288)	287	346	326	270	(296)	-	348	250	211	206	223
24	-	294	(361)	283	248	(284)	(301)	286	247	222	260	246
25	-	287	351	(295)	-	(279)	-	(331)	-	-	253	240
26	-	271	311	297	291	290	(287)	-	(265)	(228)	265	-
27	(397)	277	245	(278)	(274)	299	-	(342)	251	231	268	-
28	365	274	(222)	294	(287)	279	-	(365)	258	233	215	221
29	320		(238)	287	(267)	-	(286)	338	249	246	242	-
30	(304)		279	295	(275)	-	(287)	C	266	228	(258)	-
31	320		268		272		(268)	342		213		231

SUPPLEMENTARY TABLE 6

F2 WORLD DISTURBANCE INDEX "W"

1959

Date	Jan	Feb	Mar	Apr	May	June	July	Aug	Sept	Oct	Nov	Dec
1	1	1	5	2	4	3	2	3	2	3	3	2
2	0	2	4	2	2	3	2	3	3	3	4	2
3	0	3	4	2	2	4	1	2	2	2	3	3
4	1	2	3	0	2	3	2	2	4	3	3	2
5	2	3	3	0	3	3	3	2	4	3	3	3
6	3	3	2	1	0	2	2	2	2	4	2	2
7	3	2	2	3	1	2	3	2	1	2	2	1
8	2	0	3	3	4	2	3	2	1	1	2	0
9	3	1	2	4	2	4	2	2	0	1	1	1
10	5	1	1	5	2	3	3	3	1	0	1	1
11	4	2	1	3	3	2	3	2	1	1	1	0
12	2	3	2	2	4	3	5	1	1	1	0	1
13	1	2	2	1	3	2	3	1	0	0	1	2
14	1	3	1	2	3	2	3	1	2	2	2	3
15	1	5	1	2	3	2	4	3	2	1	1	2
16	2	5	0	2	3	2	5	5	2	0	1	2
17	3	4	0	2	2	0	4	5	2	2	1	1
18	2	2	1	2	2	1	5	4	2	2	1	1
19	1	2	1	1	2	1	4	4	2	2	2	2
20	1	0	1	1	2	0	2	3	5	1	0	1
21	1	1	2	1	3	1	2	3	5	1	1	1
22	0	1	1	1	3	2	2	2	3	3	2	1
23	0	2	2	3	3	2	1	2	2	3	3	2
24	0	1	1	5	5	3	2	1	3	2	2	2
25	2	4	3	4	5	1	3	1	3	3	1	2
26	3	5	4	3	3	3	4	1	3	3	2	2
27	2	5	5	3	2	3	3	0	2	2	1	3
28	1	4	5	4	1	4	2	0	2	1	3	2
29	2	-	5	3	1	5	1	0	1	1	2	2
30	2	-	4	3	3	5	0	1	1	2	3	2
31	2	-	2	-	3	-	2	2	-	4	-	2

SUPPLEMENTARY TABLE 7

Polar Cap Absorption Events

1959	Beginning UT	Magnitude	Db	Duration Days	Sources
May 10	2300	Very Large	> 15	7.0	B, RL, K
July 10	0400	Very Large	> 15	4.0	B, RL, K
July 14	b0700	Very Large	> 15	3.0	B, RL, K
July 16	b2250	Large	> 15	3.0	B, RL, K
Sept. 2	040	Very Small		2.0	B

B=D. K. Bailey - Effects on VHF forward scatter signals

RL=G. C. Reid and H. Leimbach - Riometer measurements of absorption of cosmic noise

K=Kiruna, Sweden, Geophysical Observatory - Riometer measurements of absorption of cosmic noise

Printed in Dunstable, United Kingdom

91221452R00065